Signals and Communication Technology

This series is devoted to fundamentals and applications of modern methods of signal processing and cutting-edge communication technologies. The main topics are information and signal theory, acoustical signal processing, image processing and multimedia systems, mobile and wireless communications, and computer and communication networks. Volumes in the series address researchers in academia and industrial R&D departments. The series is application-oriented. The level of presentation of each individual volume, however, depends on the subject and can range from practical to scientific.

Indexing: All books in "Signals and Communication Technology" are indexed by Scopus and zbMATH

For general information about this book series, comments or suggestions, please contact Mary James at mary.james@springer.com or Ramesh Nath Premnath at ramesh.premnath@springer.com.

Hamed Taherdoost

Innovating and Integrating Emerging Technologies

Springer

Hamed Taherdoost
Department of Arts, Communications and Social Sciences,
School of Arts, Science and Technology
University Canada West
Vancouver, BC, Canada

ISSN 1860-4862 ISSN 1860-4870 (electronic)
Signals and Communication Technology
ISBN 978-3-031-93952-5 ISBN 978-3-031-93953-2 (eBook)
https://doi.org/10.1007/978-3-031-93953-2

This Springer imprint is published by the registered company Springer Nature Switzerland AG
The registered company address is: Gewerbestrasse 11, 6330 Cham, Switzerland

Preface

In an era defined by rapid technological advancements, the integration and innovation of emerging technologies have become the cornerstone of progress across industries. *Innovating and Integrating Emerging Technologies* aims to provide a comprehensive exploration of the most transformative technologies shaping our present and future, offering insights into their principles and applications and the challenges they present.

As we venture deeper into the digital age, technologies such as artificial intelligence, machine learning, big data, the Internet of Things, and quantum computing are not only revolutionizing the way businesses operate but also redefining our societal structures and everyday lives. These advancements bring with them both immense opportunities and complex challenges, from fostering innovation and improving efficiency to addressing ethical considerations and navigating security risks.

This book is structured to take you on a journey through these innovations, beginning with a foundational understanding of value creation and the business models driving technological change. Each chapter delves into a specific emerging technology, presenting its core concepts, applications, and future potential. Whether it is the transformative power of cloud computing, the promise of virtual and augmented realities, or the groundbreaking possibilities of quantum computing, this book will guide readers through the nuances and implications of these innovations.

The objective is not only to understand these technologies in isolation but also to explore how they can be strategically integrated to solve complex problems and create new value. Through case studies, expert insights, and forward-looking discussions, the reader will be equipped with the knowledge necessary to navigate the fast-evolving landscape of emerging technologies.

As we continue to innovate and integrate these technologies, we must also remain vigilant about their ethical, social, and environmental impacts. From the disruptive nature of automation to the security threats posed by increased connectivity, responsible adoption and management of these technologies will be essential in ensuring their positive contribution to society.

Innovating and Integrating Emerging Technologies serves as a guide for business leaders, technologists, policymakers, and anyone interested in understanding and leveraging the potential of emerging technologies to drive sustainable innovation and growth. It is a call to action for all who seek to harness the power of innovation, to think critically about its implications, and to act with foresight and responsibility as we shape the future.

This book is a resource for those who are passionate about exploring how emerging technologies can transform industries, improve lives, and build a future that is both innovative and sustainable.

Vancouver, BC, Canada Hamed Taherdoost

Contents

About the Author

Dr. Hamed Taherdoost is an award-winning leader in research and development, with over 20 years of contributions across both industry and academia. He is the founder of Hamta Business Corporation and serves as Associate Professor and Chair of RSAC at University Canada West. Dr. Taherdoost industry experience since 2001 includes leadership roles such as CTO and R&D Manager, having worked at international companies in Cyprus, the UK, Malta, Iran, Malaysia, and Canada, contributing to projects in industries such as healthcare, transportation, residential, oil and gas, and IT. Additionally, he has served as a trusted technical and technology consultant for multiple companies, providing advisory and mentorship.

In academia, Dr. Taherdoost has held teaching positions in Southeast Asia, the Middle East, Europe and North America since 2009. He began his academic career as a lecturer at AU & PNU and later served as an adjunct professor, dissertation chair and faculty fellow at Westcliff University, USA, & adjunct professor and thesis supervisor at Gisma University of Applied Sciences, Germany. His research tenure at IAU lasted over eight years, during which he supervised numerous students.

Dr. Taherdoost has a prolific publishing record, with over 350 scientific articles in top-tier journals and conference proceedings. His work is widely recognized, with a strong citation impact reflecting in a high h-index (55), and his contributions span book chapters, edited volumes, and authored books focusing on technology and research methodology. His textbooks on E-Business and Digital Transformation are teaching at global institutions, including the University of Malta, Yarmouk University, University of Kerala, University of Macau, University Sains Islam Malaysia and University Canada West. His research accomplishments have earned him a place among the top 10 SSRN Business Authors from 2021 to present, ranked among the top 1% of researchers globally (D1 rank) based on citation metrics with 20 highly cited papers (2015–2024) according to OpenAlex and ScientoMetrics, as well as being featured in the Stanford-Elsevier list of the world's top 2% of Scientists from 2021 to present.

Dr. Taherdoost has organized and chaired many workshops and conferences and has frequently been invited as a keynote speaker. He is an active member of the editorial, reviewer, and advisory boards for several prestigious journals published by IEEE, Taylor & Francis, Springer, Emerald, Elsevier, MDPI, EAI, IGI Publishing, and Inderscience. He has also participated as an organising, scientific and technical committee member in over 400 conferences held across Europe, America, Australia, Asia, and Africa.

Dr. Taherdoost's leadership and innovation have earned him numerous accolades, including THE BIZZ Business Excellence Award, PeerJ Award, and recognition at the Asia Corporate Excellence & Sustainability Awards. He was a finalist for the Southeast Asian Startup Awards by Global Startup Awards. Additionally, his research achievements also include winning several best paper awards, outstanding reviewer awards and best presentation awards like MLIS awards, CSA award, CSTFM Lecture award, the Outstanding Editorial Board Member award from Bilingual Publishing, Best Paper of the Year of Computers MDPI, Editor's Choice Articles of Algorithm Journal, and Best Interview Award of Encyclopedia. His teaching innovations have also been recognized, including being a finalist for the Innovation in Teaching of Research Methodology Excellence Award (2022 & 2025) and winning the University Canada West's Sustained Excellence in Teaching and Learning Award.

Dr. Taherdoost is the Editor of a book series with Routledge (Taylor & Francis Group) titled Mastering Academic Excellence: Research, Teaching, Learning, and Publishing & Editor of a book series entitled "Quantum Computing" with Jenny Stanford Publishing Pte Ltd. He serves on the editorial boards of several high-impact journals, including the International Journal of Information Technology Project Management (IGI, IF: 0.8), Transactions on Scalable Information Systems (EAI, Q2), International Journal of Electronic Government Research (IGI, IF: 1.2), Information Resources Management Journal (IGI, IF: 1.4), Journal of Blockchain (MDPI), and International Journal of Data Mining, Modelling, and Management (InderScience, IF: 0.5, CS: 0.9). He is also an Associate Editor of Frontiers in Research Metrics and Analytics (Scopus) and Academic Editor of PLOS ONE, Editor of Discover Computing (Springer), and is Area Editor of the IEEE Canadian Journal of Electrical and Computer Engineering (CS:3.7 & IF:2.1). He is a Certified Cyber Security Professional and Certified Graduate Technologist. He is a GUS Fellow - GUS Institute | Global University Systems, senior member of IEEE, and Working Group Member of International Federation for Information Processing - IFIP TC 11 - Human Aspects of Information Security and Assurance and Information Security Management. Currently, he is involved in several multidisciplinary research projects, including studying innovation in information technology, blockchain, cybersecurity, and technology acceptance.

Chapter 1
Innovation and Value Creation

1.1 Introduction

Innovation, the act of turning ideas into products or services, is crucial to value generation. This process introduces new technology and improves products, services, and business models. According to Colombo et al. [1], firms need to balance innovation and value creation to succeed in a complex and dynamic market. Companies that use innovation can differentiate themselves, fulfill changing client wants, and seize new market opportunities, improving profitability and sustainability.

Value creation includes innovation's influence on society and the environment, not just financial measures. Emerging technologies may solve climate change and socioeconomic inequity. New technologies can create economic, environmental, and social value, supporting the UN Sustainable Development Goals (SDGs) [2]. Sustainability and corporate social responsibility are stressed in this comprehensive approach to value creation, encouraging organizations to innovate for the benefit of their bottom line and the communities they serve.

The link between innovation and value development is complicated. In open innovation situations where information sharing is crucial, organizations typically struggle to create and capture value. Open innovation enterprises need to combine collaborative value generation and proprietary knowledge protection [3]. This contradiction underlines the need for effective techniques that allow firms to harness external ideas and resources while protecting own IP. Navigating these challenges can boost innovation and competitiveness.

Entrepreneurship is crucial to innovation and value generation. Innovative ideas often sprout in entrepreneurial ventures, challenging standard corporate norms. Entrepreneurial marketing tactics, which discover and meet client requirements through new methods, create value [4]. Entrepreneurship and innovation create a

H. Taherdoost, *Innovating and Integrating Emerging Technologies*, Signals and Communication Technology, https://doi.org/10.1007/978-3-031-93953-2_1

culture of creativity and agility, allowing organizations to adapt to market changes and capitalize on new trends.

The essential link between creative ideas and the creation of value inside companies is investigated in this chapter. It explores the nature and kinds of innovation and shows how companies may support and pay for these projects from many sources. Emphasizing the need of creating and adjusting corporate models to maintain competitive advantage, the chapter also looks at the business model theory. It outlines what value is and shows how companies could generate, present, and evaluate it successfully. By use of impact assessment, the chapter assesses the wider consequences of creativity and offers a road map for negotiating the complexity of implementing novel ideas into reality.

1.2 Innovation

Innovation is the development and application of fresh ideas, methods, goods, or services that noticeably increase results, efficiency, effectiveness, or quality. It is different from simple creativity, which is the production of fresh ideas without necessarily application. Thus, innovation is the useful implementation of artistic ideas to produce value.

Product, process, and business model improvements all count toward the multifarious definition of innovation that is essential for corporate success (Fig. 1.1). As shown by the revolutionary influence of smartphones, product innovation concentrates on creating new or greatly enhanced goods or services, upgrading features,

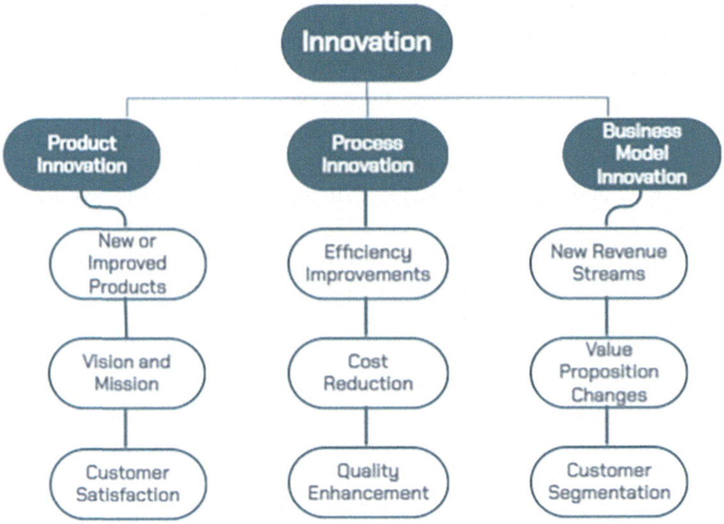

Fig. 1.1 Product, process, and business model innovation

functionality, or performance to suit changing customer wants and grab new market opportunities. Through developments in manufacturing techniques, technologies, or workflows, such as lean manufacturing approaches that simplify processes and raise productivity, process innovation seeks to improve efficiency, lower costs, and raise quality. Conversely, business model innovation, best shown by disruptive models, involves basic transformations in how a company generates, delivers, and captures value, therefore changing revenue sources, customer connections, or value offerings. Although every kind of innovation aims at distinct facets of a company, they are related and usually motivate and improve one another. To fit changes in the market, increase operational efficiency, and provide more value to consumers, a good innovation strategy combines product, process, and business model improvements, thereby guaranteeing sustained development and competitiveness.

Innovation enables companies to expand. Though expressed in turnover and profit, growth also happens in knowledge, experience, efficiency, and quality. Applied to goods, services, or processes, innovation, which can be radical or incremental, is the process of introducing changes to already existing. From departments to management teams to personally, it can occur at all levels. Many elements inspire and propel a company to be creative. Every one of these forces depends on continuity and education. These forces inspire urgency to develop fresh organizational objectives and produce fresh ideas for achieving them. Though it can also refer to processes that produce goods or services including intangibles, the word innovation is usually connected with products [5]. Many obstacles can hinder the process of innovation even with the motivating factors. Figure 1.2 shows the main elements causing both encouragement and hindrance of invention.

1.3 How to Finance and Fund Innovation

Business development and competitive advantage depend on innovation; yet, financing it often proves difficult. Using retained revenues or specialized company assets set aside for research and development, companies frequently encourage innovation via internal finance. While corporate funds provide a fixed budget consistent with strategic aims, retained earnings let businesses invest revenues free from outside limitations. These tools can be constrained, though, and they could not be sufficient for major projects, which calls for other financing sources.

Though they provide large resources, outside funding sources including venture capital, private equity, and crowdsourcing have trade-offs. In exchange for stock, venture capitalists fund early-stage, high-risk businesses, therefore offering not only funding but also useful knowledge and contacts. Investing in more established businesses or those seeking growth, private equity firms provide major funding and strategic direction in exchange for a sizable equity ownership share. Crowdfunding offers a chance to get money without sacrificing equity and is a marketing tool since it generates little sums from many people. Notwithstanding these benefits, outside

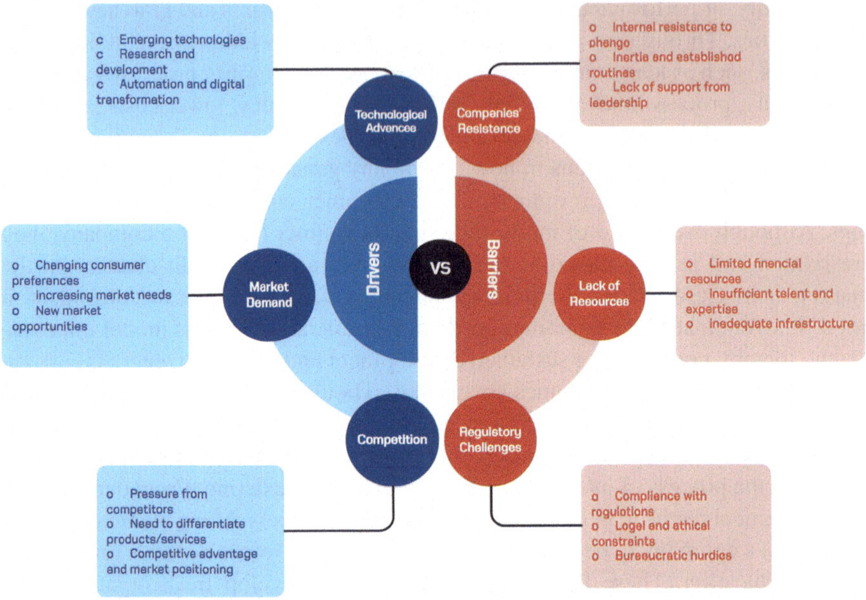

Fig. 1.2 Drivers and barriers to innovation

financing sources can call for successful promotion, complicated negotiations, or diluting of equity.

Two further sources of support for innovation are public money and strategic alliances. Though they come with competitive application procedures and compliance requirements, government grants and tax incentives give non-repayable funding or tax relief to spur research and development. Strategic alliances with colleges, research labs, or industrial consortia let companies pool resources, lower risks, and access specialized knowledge. Though often difficult, these partnerships provide useful tools and support. Another type of outside investment is corporate venture capital, in which big companies fund startups or creative ideas inside connected businesses in line with strategic objectives and maybe include fresh ideas into existing operations.

Supporting the creation and commercialization of innovative ideas, the phases of innovation funding are absolutely vital (Fig. 1.3). Usually supplied by angel investors, early-stage venture capital, or crowdsourcing sites, funding is employed in the seed stage to turn an initial idea into a viable good or service. This phase concentrates mostly on verifying the concept and creating prototypes. Turning now to the early stage, money is meant for operations scaling, market research, and product improvement. At this point, investors include venture capitalists ready to assume more risks for significant stock interests. Funding helps a company scale up, reach more markets, and raise production capacity throughout its growing phase. Often this entails private equity or later-stage venture financing, which offers more funding to hasten expansion and take market share. Every funding level fits the

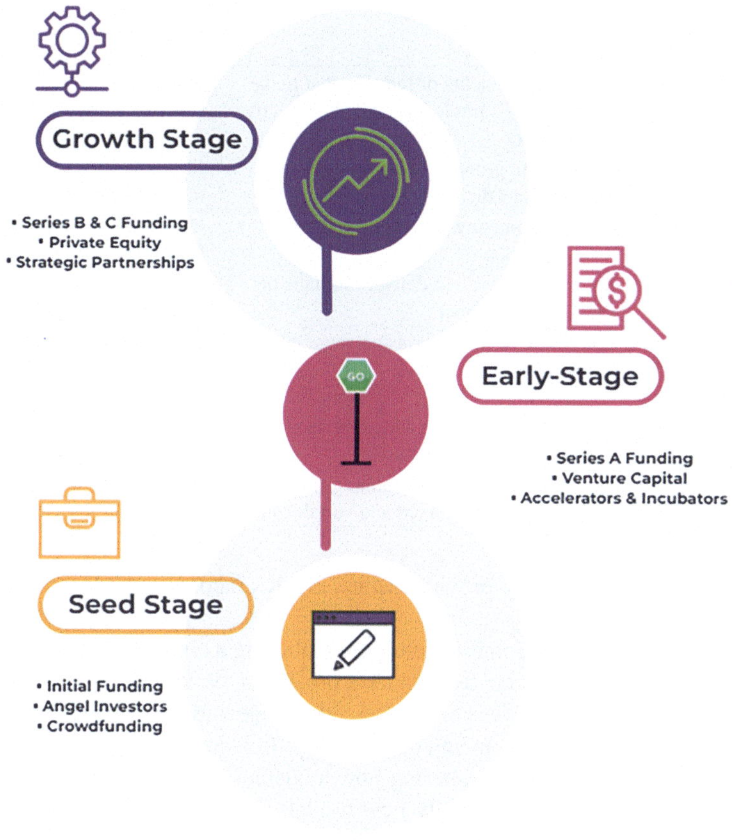

Fig. 1.3 Stages of innovation funding

development demands of the business and growth trajectory, therefore enabling the path from idea to market leader.

1.4 The Business Model Theory

A business model is really a story about how a company runs and makes money. Emphasizing how value is generated for consumers and how that value is then acquired by the company itself, it covers the policies and techniques that direct a company's contacts with its market. A well-defined business model is important because it guarantees that all stakeholders know how the company wants to succeed in its market environment [6], therefore clarifying the operational mechanics of a company. Table 1.1 lists the basic elements forming a business model. Understanding

Table 1.1 Core elements of a business model

Element	Description	Examples
Value proposition	The unique benefits or features that make the company's product or service attractive to customers	Innovative technology, superior quality, cost savings
Customer segments	The specific groups of people or organizations the business aims to serve	Millennials, small businesses, luxury consumers
Revenue model	How the company generates income	Direct sales, subscription fees, licensing, advertising
Distribution channels	The methods used to deliver products or services to customers	Online stores, physical retail locations, partner networks
Customer relationships	The nature of interactions between the business and its customers	Customer service, loyalty programs, personalized support
Key resources and activities	The critical assets and activities necessary to create and deliver value	Intellectual property, manufacturing facilities, R&D
Cost structure	The financial aspects of running the business, including fixed and variable costs	Salaries, raw materials, marketing expenses

the operations of a company provides value, creates income, and depends on each component.

Business model innovation is developing or altering a company's business model to improve performance, seize fresh market prospects, or adapt to evolving circumstances. This innovation drives competitive advantage and development by unlocking new revenue sources, enhancing operational efficiency, or generating distinctive consumer experiences, thereby altering how a business interacts with its surroundings. This can entail creating totally new models, such platform-based models or freemium models, or modifying existing models, including switching to subscription-based income or reconfiguring distribution channels.

1.5 What Is Value?

In the context of a company, values often relate to the basic ideas and beliefs that direct decision-making and organizational behavior. From strategic direction to corporate culture, these principles can affect everything and help to define how a business interacts with its stakeholders and the market. Indicating what is thought excellent or desirable inside the company, they act as both descriptive and normative ideas [7]. Value is sometimes connected in sustainable business models to the long-term effects of company activities on the environment and society. From this point of view, especially in relation to sustainability and ethical issues, clarity in defining what qualifies as value is of great importance. The difficulty to express value in this case parallels more general difficulties matching corporate objectives with society demands [8].

Understanding business dynamics requires first a grasp of the ideas of value generation and value appropriation. Value creation is the process by which companies provide advantages for consumers; value appropriation is the process by which some of that produced value is captured by companies for themselves. Strategic management and competitive advantage both depend on this duality [9].

Strategic aim and management philosophy of a company depend on its corporate values. Along with general financial performance, they can greatly affect customer satisfaction and loyalty. Declared values and real-world behavior might be in conflict that compromises company performance and stakeholder confidence [10]. Value propositions in marketing help to define how a company plans to satisfy its consumers. This is conveying the advantages and special offers that set a company apart from its rivals. Value propositions are therefore a crucial part of business strategy since their efficacy can affect consumer impressions and buying choices [11].

The debate revolves mostly on two main points of view: shareholder value and customer value (Fig. 1.4). Developing good company plans that propel sustainable success depends on an awareness of the differences between these two types of value. Studies show that more shareholder value may follow from more customer pleasure. Mostly, this relationship is mediated by lower cash flow volatility and increased cash flow growth. More likely to show loyalty, repeat business, and good word-of-mouth are satisfied consumers, which taken together help to create a more consistent and predictable income source. One study, for example, revealed that customer happiness directly increases shareholder value by improving cash flow growth and so reducing its volatility, thereby improving general financial performance [12, 13].

Linking customer value to shareholder value depends mostly on the idea of Customer Lifetime Value (CLV). Over the course of a commercial partnership, CLV estimates the overall income a single client account can bring. Treating clients as assets helps companies to strategically control client relationships, thereby optimizing CLV, which increases shareholder value. The approach proposes that evaluations of investments in customer connections should be focused on their possibility to improve future cash flows, therefore influencing shareholder returns favorably [12, 14].

Strategically, businesses that give customer happiness and engagement first priority will probably find their market value to improve. This calls for a change in how companies gauge success—moving beyond conventional financial criteria to include customer-centric assessments that mirror long-term value creation. The integration of marketing tactics that increase customer happiness shows that customer value and shareholder value are not mutually exclusive but rather linked [14, 15], hence improving financial results.

Both very vital for success and sustainability, values can also be classified as concrete and immaterial (Table 1.2). Although concrete values offer necessary basic support for operations, intangible values are becoming more and more important for creativity, competitive advantage, and general company performance. Companies who successfully combine these kinds of resources are more suited to withstand long-term success and change with the times [16, 17].

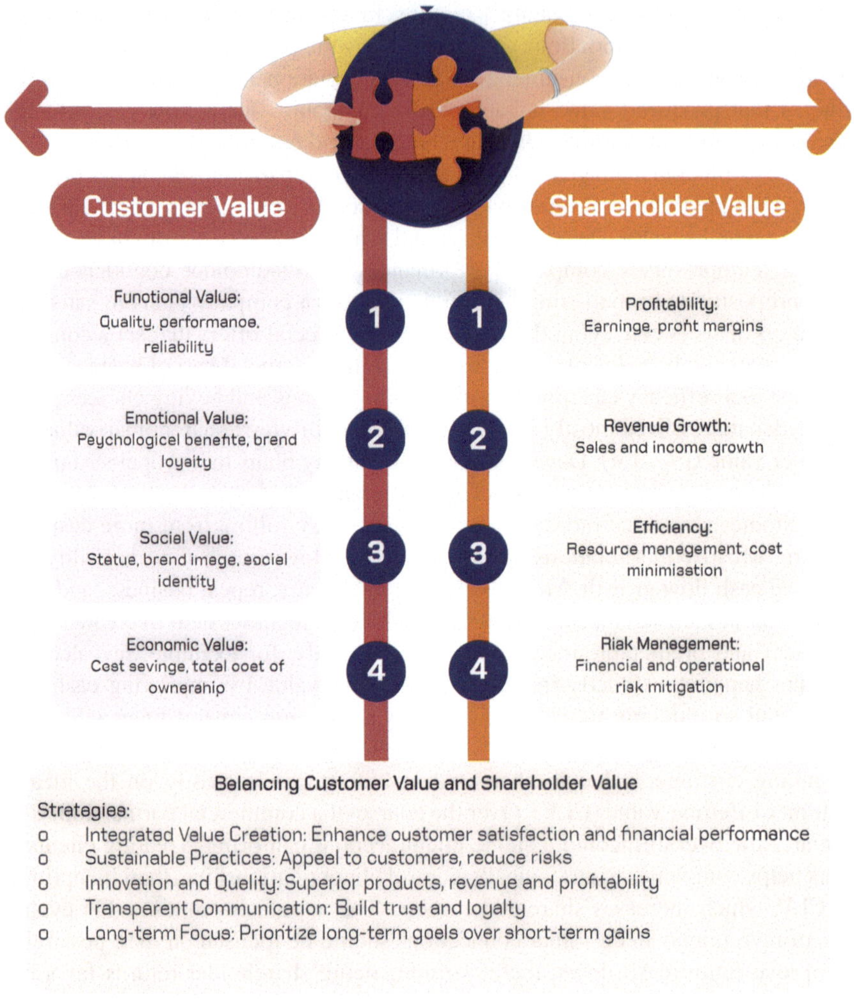

Fig. 1.4 Customer value vs. shareholder value

Tangible value, that is, the physical, quantifiable features of a good or service, such as quality, price, and convenience, that directly affect a company's bottom line and are more easily expressed with hard data. One could cite the high-resolution camera of a smartphone or the cost-cutting features of a program solution. But intangible worth includes non-physical elements like brand reputation, customer experience, and innovation that greatly affect consumer loyalty and satisfaction. Often shown in increased customer lifetime value and a better market position, strong brand reputation, great customer service, and innovative technology demonstrate intangible worth. While tangible value stimulates instant sales, intangible value promotes long-term loyalty and competitive advantage and need to be

Table 1.2 Comparison of tangible and intangible value

Aspect	Tangible value	Intangible value
Definition	Physical and measurable benefits	Non-physical and subjective benefits
Measurability	Easily quantifiable (e.g., price, features)	Often subjective and harder to quantify
Examples	Product features, price, convenience	Brand reputation, customer experience, innovation
Impact on business	Directly affects immediate sales and financial performance	Influences customer loyalty, long-term brand strength
Communication	Can be demonstrated with concrete evidence (e.g., product specs)	Requires more nuanced communication (e.g., storytelling, branding)
Assessment methods	Performance metrics, customer reviews, cost-benefit analysis	Customer surveys, brand equity measurement, sentiment analysis
Investment focus	Enhancing product quality, optimizing pricing, improving convenience	Building brand, enhancing customer service, fostering innovation
Customer perception	Direct and straightforward (e.g., high-quality materials)	Subtle and emotional (e.g., brand trust, user experience)
Long-term impact	Often provides immediate, short-term gains	Crucial for long-term loyalty and competitive advantage
Examples of companies	Companies offering high-quality, reliable products	Companies with strong brand identities, exceptional customer service

balanced with other forms of value. Businesses should be always innovating, strengthening their brands, emphasizing customer experience, and using intellectual property to maximize both kinds of value, so presenting a strong proposal that draws in and keeps consumers, thus guaranteeing sustainable development and market leadership.

Value creation is fundamentally about creating goods or services that satisfy consumer requirements and wants in a way that offers notable advantages over current substitutes. Understanding consumer demands through research and feedback starts this process; then, creation and delivery of solutions that efficiently meet those needs follow. From manufacturing and distribution to marketing and sales, value delivery covers the whole process of introducing the value proposition to the market. It guarantees that the produced value is affordable, timely, and easily reachable for the targeted consumers.

Understanding how effectively a company is achieving its objectives and providing advantages to its stakeholders depends on knowing its value. Value shows itself in many different forms, financial returns, consumer pleasure, brand reputation, and so on, both real and intangible. Correct measurement lets companies show their influence, maximize their plans, and make wise judgments. Table 1.3 offers a summary of main performance indicators (KPIs), ROI computation, cost-benefit analysis, and various impact measuring techniques applied to assess financial, operational, customer, and environmental value inside a company.

Table 1.3 Key metrics and methods for measuring business value and impact

Category	Metric/method	Description	Interpretation
Financial KPIs	Revenue growth	Measures the increase in sales over a period	Higher growth indicates increasing sales performance
	Profit margins	Shows the percentage of revenue that exceeds total expenses	Higher margin suggests better profitability
	Return on investment (ROI)	Assesses the profitability of an investment	Higher ROI indicates a more effective investment
Operational KPIs	Efficiency ratios	Measures how efficiently resources are used	Lower ratios indicate better efficiency
	Quality scores	Evaluates the quality of products or services	Higher scores reflect better quality
	Time-to-market	Time taken to develop a product and bring it to market	Shorter time indicates faster development and market entry
Customer KPIs	Customer satisfaction scores (CSAT)	Measures customers' satisfaction with products or services	Higher scores reflect greater customer satisfaction
	Net promoter score (NPS)	Gauges customer loyalty and likelihood to recommend	Higher scores indicate stronger customer loyalty
	Customer lifetime value (CLV)	Predicts the total revenue a business can expect from a customer over their lifetime	Higher CLV signifies more valuable customers
Cost-benefit analysis	Purpose	Compares total expected costs to total expected benefits	Positive net benefit indicates a worthwhile investment
Measuring impact	Social impact	Assesses contributions to societal well-being, such as job creation and community development	Positive impact signifies beneficial societal contributions
	Economic impact	Evaluates the contribution to economic growth, revenue generation, and cost savings	Positive impact indicates economic benefits and development
	Environmental impact	Measures effects on the environment, such as resource usage and emissions	Lower impact reflects better environmental sustainability

1.6 Impact Assessment

Understanding the efficacy, efficiency, and general results of an effort is the main focus of impact evaluation. Methodically examining these effects helps companies to make wise decisions, strengthen their policies, and show stakeholders responsibility. It helps ascertain whether an invention has met its intended purposes. It assesses if the new good, service, or process has achieved the promised advantages, that is, higher productivity, more customer satisfaction, or better operational

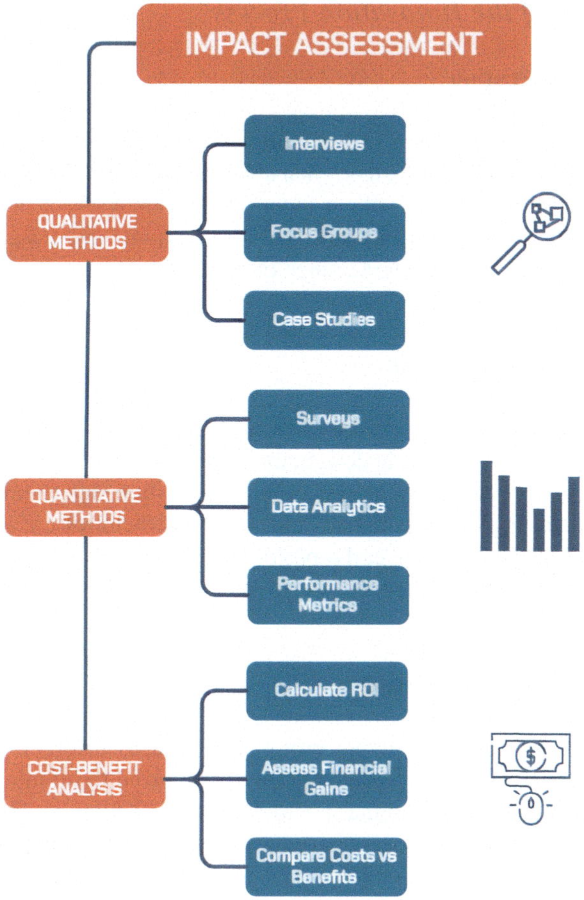

Fig. 1.5 Methods of impact assessment

efficiency. Impact evaluation looks for more general consequences than only financial rewards. This covers social effects including employment creation and neighborhood well-being; economic effects including market expansion and cost savings; and environmental effects including sustainability and resource preservation. A flowchart in Fig. 1.5 graphically arranges the several approaches used for impact assessment into three primary groups.

1.7 The Innovation Journey

Innovation is a journey with multiple stages, each with unique opportunities and obstacles; it is not a linear process. Organizations looking to launch new concepts and achieve sustained growth need to comprehend and navigate this process.

1.7.1 Stages of the Innovation Process

1.7.1.1 Idea Generation

Idea generation is the first step in the innovation process. This phase entails the creative process of coming up with original concepts and fixes. It may be sparked by internal brainstorming sessions, client input, market analysis, or developments in technology. Methods like ideation workshops, hackathons, and design thinking can foster creativity and generate a wide range of concepts.

1.7.1.2 Concept Development

After ideas are produced, they need to be refined into workable concepts. In order to create a cohesive and workable design, the basic concepts need to be refined and expanded upon at this stage. Defining the value proposition, determining target markets, and summarizing the possible advantages and difficulties are all part of concept development. Developing early prototypes or models during this stage is common to test and validate the idea.

1.7.1.3 Prototyping and Testing

The practice of developing concrete prototypes of ideas in order to assess their viability and functionality is known as prototyping. From basic mock-ups to fully functional models, prototypes come in various forms. Testing entails getting input from stakeholders or potential users in order to evaluate the concept's viability and pinpoint areas in need of development. To make sure the finished product satisfies the requirements and expectations of its intended audience, iterative testing and refinement are essential.

1.7.1.4 Commercialization

The refined product or service is introduced to the market during the commercialization stage. This encompasses the completion of production processes, the development of marketing strategies, and the establishment of distribution channels. Careful planning and coordination across a variety of departments, such as marketing, sales, and operations, are necessary for commercialization as well. It also entails the expansion of the innovation to a broader audience and the attainment of market penetration.

1.7.2 *Managing the Innovation Journey*

The literature underlines the need of building a creative and cooperative organizational culture by means of innovation. Building an atmosphere fit for innovation depends mostly on good leadership, a common vision, and suitable organizational structures. Companies that include staff members at all levels of the innovation process typically produce superior results [18]. This interactive strategy not only improves staff involvement but also makes use of several points of view, so producing more creative ideas. Managing the innovation process ultimately calls for a thorough knowledge of the organizational dynamics supporting effective innovation as well as the involved procedures.

Managing the innovation process means methodically encouraging creativity and putting fresh ideas into use inside companies. This process consists in multiple phases: concept generation, development, and commercialization among others. Managing this road depends on knowing how technical developments interact with market needs, which calls for companies to stay flexible and changeable. Co-creation and user involvement need to be seamlessly combined since they improve the innovation process by matching goods and services to consumer demands. This cooperative strategy not only raises the internal innovation rate but also helps companies keep a competitive edge in fast changing markets [18].

Guiding companies across their innovative paths is heavily influenced by strategic niche management. Organizations can develop and polish new technologies free from the immediate demands of market rivalry by establishing protected areas for research. This approach lets one investigate radical ideas progressively merging them into current systems. Effective management of innovation also calls for a cultural transformation inside companies, thereby fostering an attitude that welcomes change and encourages ongoing education. Creating such an atmosphere helps companies to successfully negotiate the complexity of innovation and realize sustainable development [19].

1.8 Conclusion

This chapter examined the complex relationship between innovation and value creation and their importance in current corporate performance. Internal finance and strategic collaborations promote innovation from modest advances to radical breakthroughs. Companies may turn innovative ideas into marketable products and services with effective finance methods, innovation lifecycle alignment, and business model theoretical knowledge. Companies demonstrate that innovation goes beyond new products to changing business strategies to meet market and customer needs. Effective measurement of actual and intangible value generation for customers and shareholders is essential for competitive advantage. Impact assessment evaluates innovation's social, economic, and environmental impacts using thorough data gathering and analysis. Innovation demands strategic planning, strong leadership, and a supportive culture from ideation to commercialization. In the future, new technology, consumer tastes, and global issues will affect innovation and value generation.

References

1. Colombo, M.G., et al.: Open innovation, value creation and value capture: an introduction. J. Ind. Bus. Econ. **50**(4), 731–742 (2023)
2. Küfeoğlu, S.: Innovation, value creation and impact assessment. In: Emerging Technologies, pp. 1–40. Springer, Cham (2022)
3. Toroslu, A., et al.: Value capture in open innovation: a literature review and a research agenda. Ind. Mark. Manag. **114**, 297–312 (2023)
4. Risitano, M., et al.: Analysing the relevance of value creation in the interconnection amongst entrepreneurship, marketing and innovation: a systematic literature review. Int. J. Entrep. Behav. Res. (2023). https://doi.org/10.1108/ijebr-02-2022-0203
5. Baporikar, N.: Drivers of innovation. In: Knowledge Management for Competitive Advantage During Economic Crisis, pp. 250–270. IGI Global, Pennsylvania, PA (2015)
6. Kaplan, S.: Business models 101: Creating, delivering, and capturing value. In: The Business Model Innovation Factory, pp. 17–33. Wiley, Hoboken, NJ (2012)
7. Askeland, H.: Values—reviewing the construct and drawing implications for values work in organisation and leadership. In: Understanding Values Work: Institutional Perspectives in Organizations and Leadership, pp. 15–34. Springer, Cham (2020)
8. Neesham, C., Dembek, K., Benkert, J.: Defining value in sustainable business models. Bus. Soc. **62**(7), 1378–1419 (2023)
9. Hallberg, N.L.: What is value and how is it managed? J. Creat. value. **3**(2), 173–183 (2017)
10. Taher, A.: Do corporate values have value? The impact of corporate values on financial performance. Fut. Bus. J. **9**(1), 76 (2023)
11. Eggert, A., et al.: Conceptualizing and communicating value in business markets: from value in exchange to value in use. Ind. Mark. Manag. **69**, 80–90 (2018)
12. Anderson, E.W., Fornell, C., Mazvancheryl, S.K.: Customer satisfaction and shareholder value. J. Mark. **68**(4), 172–185 (2004)
13. Gruca, T.S., Rego, L.L.: Customer satisfaction, cash flow, and shareholder value. J. Mark. **69**(3), 115–130 (2005)
14. Stahl, H.K., Matzler, K., Hinterhuber, H.H.: Linking customer lifetime value with shareholder value. Ind. Mark. Manag. **32**(4), 267–279 (2003)
15. Matzler, K., et al.: The relationship between customer satisfaction and shareholder value. Total Qual. Manag. Bus. Excell. **16**(5), 671–680 (2005)
16. Jancenelle, V.E.: Tangible− Intangible resource composition and firm success. Technovation. **108**, 102337 (2021)
17. Kamasak, R.: The contribution of tangible and intangible resources, and capabilities to a firm's profitability and market performance. Eur. J. Manag. Bus. Econ. **26**(2), 252–275 (2017)
18. Tidd, J., Bessant, J.R.: Managing innovation: integrating technological, market and organizational change. John Wiley & Sons, Hoboken, NJ (2020)
19. Schot, J., Geels, F.W.: Strategic niche management and sustainable innovation journeys: theory, findings, research agenda, and policy. Dyn. Sustain. Innov. Journeys, 17–34 (2013). https://doi.org/10.1080/09537320802292651

Chapter 2
Navigating Big Data Revolution

2.1 Introduction to Big Data

Big Data is all about the massive amounts of data that are generated quickly and come in different formats from many other sources. Big Data includes a wide range of data types, including structured, semi-structured, and unstructured data, in contrast to the more structured data in traditional databases. The key features of Big Data, commonly called the Four V's, volume, velocity, variety, and veracity, embody its intricate nature and vast possibilities. Volume is all about the enormous amounts of data generated constantly, ranging from social media interactions to sensor readings. The speed at which this data is produced and needs to be processed, often in real time, is emphasized by velocity. Diversity is evident in the various data types, such as text, images, video, and sensor data. The quality and accuracy of the data can vary greatly, and it is important to consider its veracity [1].

Technological advancements and the growing digitization of society have fueled the shift from traditional databases to Big Data. Traditional databases, with their structured format and reliance on SQL (Structured Query Language), are ideal for transactional applications but face challenges when handling the vast scale and diverse nature of Big Data. The shift started with the emergence of NoSQL databases, which were created to handle large-scale and diverse data more flexibly. Technologies like Hadoop and its ecosystem have revolutionized how we handle and analyze massive datasets, enabling distributed storage and processing. This advancement has been further accelerated by the emergence of real-time processing frameworks like Apache Spark, which facilitates the swift analysis of streaming data. The significance and influence of Big Data are immense and extensive across different industries. Big Data analytics in healthcare empower personalized medicine, predictive modeling for disease outbreaks, and enhanced patient care through data-driven insights. Big Data plays a crucial role in the finance industry by enabling fraud detection, risk management, and personalized financial services. Retailers use

H. Taherdoost, *Innovating and Integrating Emerging Technologies*, Signals and Communication Technology, https://doi.org/10.1007/978-3-031-93953-2_2

Big Data to gain insights into customer behavior, improve supply chains, and provide a more personalized shopping experience through targeted marketing. The transportation industry leverages Big Data to effectively manage traffic, proactively maintain vehicles, and optimize logistics. In smart cities, Big Data enables efficient resource management, including energy consumption, waste management, and public safety. The role of Big Data goes beyond these industries, impacting scientific research, education, and government policies. For example, in scientific research, the acceleration of discoveries in genomics, climate science, and physics is greatly facilitated by large-scale data analysis. Big Data is utilized by educational institutions to enhance learning outcomes with the help of adaptive learning technologies and to streamline institutional operations. Governments extensively use Big Data to make informed public policy decisions, effectively respond to disasters, and enhance public services [2].

2.2 The Four V's of Big Data

The four V's of Big Data reflect the primary potential and challenges of managing and using enormous volumes of data (Fig. 2.1). The first V, Volume, addresses the massive amounts produced and stored every second. Driven by the rise of digital devices, social media, and IoT sensors, this data tsunami begs for scalable processing systems and better storage options. This sheer size of conventional databases presents issues for them; consequently, distributed solutions such as Hadoop and cloud storage technologies, which can efficiently handle petabytes and exabytes of data, are required.

Velocity is the second V, stressing the incredible speed at which data is created and managed. Many times, including financial transactions, social media feeds, and sensor data from smart devices, real-time or near-real-time processing is essential. This requires the acceptance of Apache Spark and Flink, robust data processing tools designed to manage and analyze streaming data on demand. Quick insights and reactions to new trends and challenges made possible by fast data processing allow businesses to adapt.

The third V, variability, addresses the many data sources and types. Big Data contains unstructured data, text, images, videos, and social media posts; semi-structured data, XML and JSON files; and structured data, traditional databases. Dealing with this diversified data environment requires flexible data processing tools to mix and assess different kinds and adjustable storage alternatives easily. Technologies such as data lakes, which preserve raw data in its native form, and NoSQL databases, which are schema-less, are key in controlling the variety inherent in Big Data.

The fourth V, veracity, tackles reliability and data accuracy. Given the massive amount of data collected from numerous sources, maintaining data quality becomes a great challenge. Incorrect judgments and bad conclusions follow from inaccurate or inadequate data. Therefore, data integrity rests mostly on solid data governance

Fig. 2.1 The four V's of Big Data

rules, including data cleaning, validation, and provenance tracking. Dealing with the veracity issue mostly relies on instruments and models that enable data quality management and provide ways of verifying and checking data.

2.3 Technologies and Tools in the Big Data Ecosystem

Figure 2.1 illustrates the four V's of Big Data, highlighting the opportunities and challenges associated with managing and utilizing large volumes of data. The letter V, Volume, represents the vast amounts of data generated and stored every second. Propelled by the advancement of digital devices, social media, and IoT sensors, this overwhelming influx of data necessitates scalable processing systems and enhanced storage solutions. The immense size of traditional databases poses a significant challenge, so distributed systems like Hadoop and cloud storage technologies are necessary. These advanced solutions can efficiently handle massive amounts of data, ranging from petabytes to exabytes.

The second V highlights the remarkable speed at which data is generated and needs to be managed: Velocity. Immediate or nearly immediate processing is crucial in various scenarios, such as financial transactions, social media feeds, and sensor data from smart devices. It is essential to implement robust data processing systems capable of handling and analyzing streaming data as needed. Two popular options for this are Apache Spark and Flink. Efficient data processing enables companies to gain insights and swiftly respond to emerging trends and challenges.

Diversity, the third V, addresses many data types and origins. Big Data encompasses many data types, including unstructured data like text, photos, videos, and social media postings, semi-structured data such as XML and JSON files, and structured data in traditional databases. Handling this diverse data landscape requires storage options and data processing tools that can easily integrate and analyze multiple data types. One important aspect of managing the diversity inherent in Big Data is utilizing various technologies. These technologies include data lakes, which store raw data in its natural form, and NoSQL databases, which do not require a fixed schema.

Data correctness and dependability are crucial regarding the fourth V, Veracity. Dealing with the vast amount of data collected from various sources poses a significant challenge in maintaining data quality. Having complete or correct data can lead to correct assessments and correct conclusions. Ensuring data integrity is crucial, and it relies heavily on robust data governance policies encompassing data cleansing, validation, and provenance tracking. Addressing the accuracy issue relies heavily on the availability of tools and frameworks that facilitate data quality management and provide methods for verifying and validating data (Fig. 2.2).

2.4 Big Data Analytics

Big Data analytics is a multifarious and sophisticated area with many methods to examine and get insightful analysis from vast databases. Descriptive, predictive, and prescriptive analytics, each with special use, support the data analysis process (Fig. 2.3). Descriptive analytics is essentially about using prior occurrences and patterns to summarize knowledge from historical data. It presents knowledge of previous occurrences through statistical approaches and data aggregating strategies. Descript analytics is used, for instance, by companies to provide reports on operational efficiency, consumer behavior, or sales success. This helps them to see trends and patterns from past data.

Conversely, predictive analytics emphasizes statistical algorithm-based future trend prediction based on past data. Predictive analytics may project future results

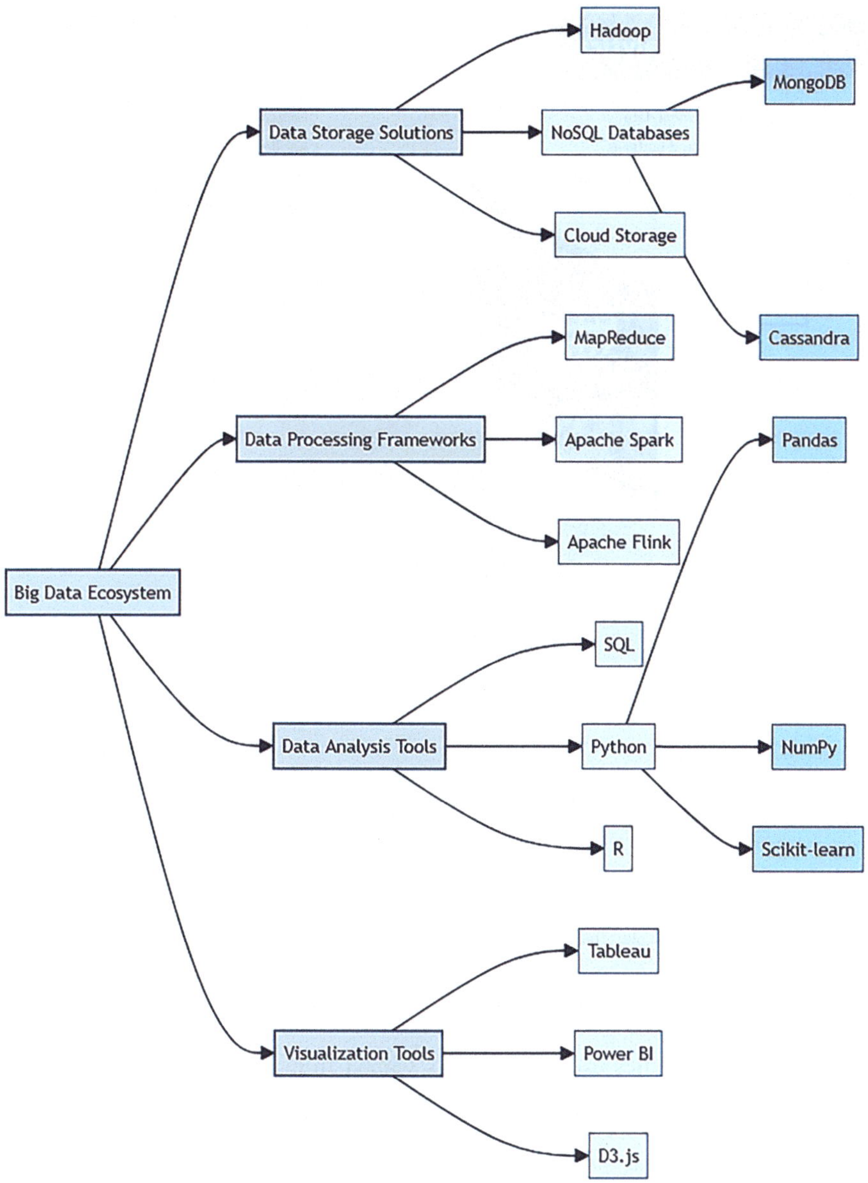

Fig. 2.2 Illustration of the Big Data ecosystem

and trends using machine learning models, time series, and regression analyses. This kind of analytics is very helpful for companies trying to make wise judgments about possible future occurrences. Predictive models, for example, may estimate consumer demand, identify prospective hazards, and project equipment breakdowns. This helps companies to act early in tackling problems and grabbing new possibilities.

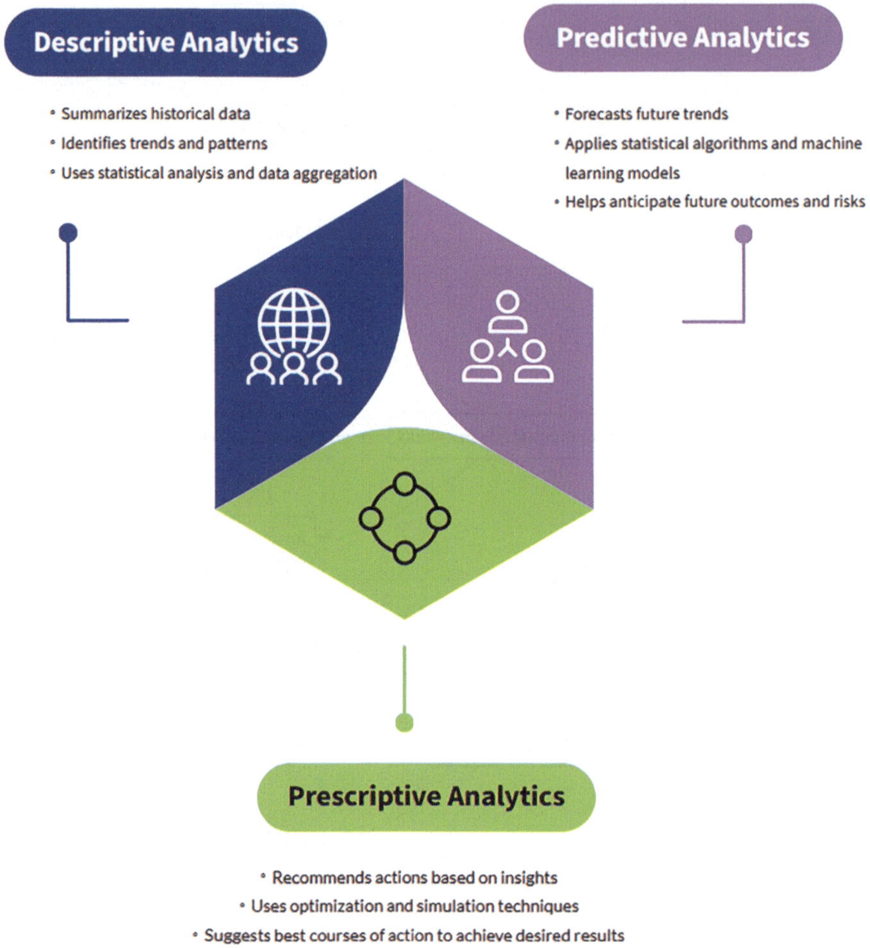

Fig. 2.3 Overview of Big Data analytics

Prescriptive analytics uses the insights gained from descriptive and predictive analytics to provide suggestions to improve outcomes, therefore transcending analysis. It suggests the best course of action using optimization algorithms, simulation tools, and decision analysis. Beyond just forecasting outcomes, prescriptive analytics offers practical actions to assist in reaching desired objectives. It may provide ideas on marketing plans to raise consumer involvement, propose ideal inventory levels to optimize sales, or advise on resource allocation to boost operational effectiveness [3].

Big Data analytics increasingly depends on artificial intelligence (AI) and machine learning (ML). These technologies provide more sophisticated data processing and analysis, enhancing conventional analytical techniques. Often exposing trends not immediately seen from conventional approaches, artificial intelligence, and machine learning algorithms can detect patterns and insights from massive amounts of information. Predictive and prescriptive analytics use machine learning models, such as decision trees, neural networks, and clustering algorithms, to deliver accurate predictions and useful suggestions. For example, systems driven by artificial intelligence may examine enormous amounts of consumer data to provide tailored suggestions, improve supply chains, and instantly spot abnormalities.

While quick decision-making and automation are made possible, integrating artificial intelligence and machine learning into Big Data analytics improves the accuracy and efficacy of analysis. These technologies let models be refined and predictions be improved with fresh data availability, enabling constant learning and adaptability. Big Data combined with artificial intelligence and machine learning offers a powerful approach for handling complex problems, simplifying company processes, and gaining a competitive edge in a society mostly dependent on data.

With its many modes of analysis, descriptive, predictive, and prescriptive, along with the developments in artificial intelligence and machine learning, Big Data analytics generally provides a broad spectrum of tools and approaches to translate unprocessable data into useful insights and strategic choices. Organizations trying to properly use their data and encourage creativity in today's fast-changing digital world depend on this analytical capacity.

2.5 Applications of Big Data

Big Data has revolutionized many sectors by offering sophisticated data analysis and decision-making power (Table 2.1). By allowing tailored medication and predictive healthcare, Big Data transforms patient care and medical research in the healthcare industry. Healthcare practitioners may customize treatment programs to fit every patient's profile by integrating and analyzing vast data from electronic health records, genetic data, wearables, and patient histories. This customized strategy reduces healthcare costs and improves patient outcomes and treatment efficacy. Driven by the abundance of data, predictive analytics finds trends and patterns from historical data to help early identification of possible health problems. Predictive models, for instance, allow one to forecast disease outbreaks, project patient admissions, and maximize hospital resource allocation, enabling a more proactive and effective healthcare delivery.

Big Data has completely changed the financial sector's risk management and fraud detection. Real-time data analytics are used by financial organizations to precisely find fraudulent activity, constantly monitor transactions, and spot any odd trends. Transaction data is analyzed using machine learning methods and data mining tools to find and highlight questionable activity. This lessens the possibility of financial loss. Furthermore, using market trend research, consumer behavior, and historical data, predictive analytics is indispensable in helping financial institutions manage risks. This helps with better decisions on loan approvals, investments, and

Table 2.1 Applications and impacts of Big Data across various sectors

Sector	Applications	Impact
Healthcare	– Personalized medicine – Predictive healthcare – Early disease detection	– Customizes treatment plans – Improves patient outcomes – Optimizes resource allocation – Enhances proactive care
Finance	– Fraud detection – Risk management – Real-time transaction monitoring – Predictive analytics	– Identifies fraudulent activities – Mitigates financial risks – Enhances investment decisions, improves credit management
Retail	– Customer insights – Personalized marketing – Inventory management – Supply chain optimization	– Increases customer satisfaction – Boosts sales and revenue – Reduces operational costs – Enhances product availability
Smart cities	– Traffic management – Resource optimization – Sustainable practices – Public safety	– Reduces congestion – Improves transportation efficiency – Optimizes energy and water use – Enhances urban living quality

risk-reducing techniques. Using Big Data, financial companies may enhance their power to withstand financial stability and change with the times.

Big Data has also clearly benefited the retail industry. Data analytics helps retailers be completely aware of consumer preferences and behavior. Retailers may create marketing campaigns customized to fit individual tastes and provide unique shopping experiences by studying purchase history, browsing behaviors, and social media interactions. This raises consumer loyalty and happiness and increases sales and income. Big Data also improves supply chain management using demand trend forecasts, inventory control, and logistical simplification. Increased operational efficiency, cost cutting, and better product availability follow this, boosting the whole customer experience.

Big Data is crucial for enhancing urban life in smart cities through resource allocation and optimization of traffic control. Urban areas compile information from many sources, including cameras, traffic sensors, and GPS devices, to examine traffic patterns, clear congestion, and improve public transit routes. This analytical technique helps maximize transportation systems, lower travel times, improve road safety, and minimize traffic congestion. Big Data also helps cities maximize their resources, including water and electricity. Using the study of use patterns and environmental data, cities have chances to adopt more sustainable practices, improve resource management, and address issues related to waste and energy consumption.

2.6 Challenges and Ethical Considerations

To effectively manage and use large datasets, companies must address many ethical questions and issues that Big Data brings. Given the delicate nature of the current data, maintaining information privacy and security is critical. Organizations must provide robust protections to guarantee the security and confidentiality of this information, considering the huge amount of personal and private data being gathered. Modern encryption technologies must ensure the highest data storage security and follow rigorous data security rules such as the California Consumer Privacy Act (CCPA) and the General Data Protection Regulation (GDPR) [4]. Dealing with user permission and ensuring that data-collecting methods are open and respectful of personal privacy rights constitute other privacy issues.

One should give the ethical consequences of using Big Data great thought. Big Data analytics has great power as it may help influence choices that significantly affect people and communities, generating legitimate issues about fairness and prejudice. Algorithmic biases may arise when machine learning models are taught on data reflecting past injustices or biased behaviors. Discriminatory results could follow from improper credit scoring or biased employment policies. Therefore, organizations must prioritize ethical, inclusive, and open data practices. They should aggressively find and fix any algorithms and data prejudices.

Data quality and governance management are among the most important issues that call for great attention. The data quality strongly influences the accuracy and dependability of conclusions obtained by Big Data. With issues like errors, inconsistencies, and incompleteness, good data quality may lead to accurate findings and analysis. Guaranteeing data integrity and correctness depend on implementing robust data governance systems. Establishing well-defined data management rules, maintaining data standards, and routinely assessing data quality are all part of this. Establishing defined roles and duties for data stewardship and ensuring that data is freely available and well-documented for auditing needs also constitute part of effective data governance.

In the Big Data terrain, tackling organizational and technological difficulties is vital. Managing the complexities of data integration, making sure data processing systems can manage vast volumes of data, and using advanced analytical tools are some of the difficulties in this subject. To properly control and evaluate vast volumes of data, companies must prioritize investment in a suitable technological stack and infrastructure. Organizational aspects, including Big Data projects, usually require corporate practices and culture changes. This might call for staff member training, promoting departmental collaboration and excellence, and ensuring Big Data plans complement the general corporate objectives. It is essential for companies to implement good change management strategies to leverage its benefits and effectively negotiate the always-shifting Big Data terrain.

2.7 Future Trends in Big Data

Many important themes in the next years will have a major influence on the direction of Big Data and its use. The Internet of Things (IoT) and edge computing point to a significant change in data collecting and analysis methodology. Edge computing, instead of depending on centralized cloud servers, concentrates on processing data near its source, such as on sensors or IoT devices. This method avoids delays, maximizes data in real-time processing, and increases the system's overall efficiency. Finding effective edge computing solutions becomes more important than ever, given IoT devices' explosive growth and the consequent network edge data flow. This trend is expected to drive developments in data processing, enabling more agile applications and helping intelligent systems flourish in many sectors.

Another trend taking hold is the mix of Big Data and blockchain technologies. Blockchain's distributed and secure character helps it address many data management issues. Through a clear, tamper-proof record of data exchanges, blockchain technology enhances data integrity and security. More accurate and trustworthy analytics, improved data traceability, and data sharing security are outcomes of this connection. Supply chain management, financial transactions, and data governance will advance significantly as businesses investigate the possibilities of merging blockchain with Big Data.

Additionally, on the horizon and ready to revolutionize Big Data analytics are fascinating advancements in quantum computing. Using the great power of quantum physics, quantum computing performs complex computations at rates faster than conventional computers. This significantly accelerates data processing and analysis, enabling answers to issues that are now difficult to address, given computing constraints. Though still in its early years, quantum computing has immense promise to enhance data analysis, cryptography, and optimization issues significantly. Quantum technology is predicted to be more important in the Big Data scene as it develops, enabling significant improvements in data processing capability.

Big Data professions and skills' future are emerging with these technical developments. Big Data's emergence in companies has created a need for personnel knowledgeable in data science, analytics, and allied disciplines. Your value will be much enhanced by knowing machine learning, artificial intelligence, and data engineering, as these fields are fundamental for the evolution and application of sophisticated analytics solutions. Furthermore, when new technologies such as edge computing, blockchain, and quantum computing become more popular, there will be a need for professionals with thorough knowledge of these developments who can easily integrate them into Big Data systems. Professionals must welcome lifelong learning and adaptation to stay relevant in today's always-changing technology scene.

2.8 Conclusion

Looking forward, numerous big developments will greatly affect Big Data and its uses. The Internet of Things (IoT) and edge computing point to a significant change in data collecting and analysis methodology. Edge computing, instead of depending on centralized cloud servers, concentrates on processing data near its source, such as on sensors or IoT devices. This method maximizes reaction times, increases data processing speed, and raises the system's general efficiency. Finding effective edge computing solutions is more critical than ever, given the fast expansion of IoT devices and the consequent explosion of data at the edge of networks. This trend is expected to drive developments in data processing, enable more agile applications, and help intelligent systems proliferate across many sectors.

Another trend taking hold is the mix of Big Data and blockchain technologies. Blockchain's distributed and secure character helps it address many data management issues. Through a clear, tamper-proof record of data exchanges, blockchain technology enhances the security and integrity of data. This connection presents chances to improve data traceability, increase data sharing, and enable more consistent and verifiable analytics. Blockchain and Big Data will enable companies to make major strides in data governance, financial transactions, and supply chain management.

Additionally, on the horizon and ready to revolutionize Big Data analytics are fascinating advancements in quantum computing. Using the amazing ideas of quantum physics, quantum computing performs complex computations at amazing rates faster than conventional computers could allow. This significantly accelerates data processing and analysis, enabling answers to issues that are now difficult to address, given computing constraints. Though still in its early years, quantum computing has immense promise to enhance data analysis, cryptography, and optimization issues significantly. Quantum technologies should become more important in the Big Data as they develop, improving data processing capacity.

Along with these technical developments, the professions and skills of Big Data are shifting. Big Data's emergence in companies has created a need for personnel knowledgeable in data science, analytics, and allied disciplines. Your worth will be much enhanced by knowing machine learning, artificial intelligence, and data engineering, as these fields are important for the evolution and application of sophisticated analytics solutions. Furthermore, when new technologies such as edge computing, blockchain, and quantum computing become more popular, there will be a need for professionals with thorough knowledge of these developments who can easily integrate them into Big Data systems. Professionals must keep learning and adjusting to be relevant in a fast-changing technology environment.

References

1. Rubinfeld, D.L., Gal, M.S.: Access barriers to big data. Ariz. L. Rev. **59**, 339 (2017)
2. Harrison, G.: Next Generation Databases: NoSQLand Big Data. Apress, New York (2015)
3. Jubi, R.: Business Analytics-Unleashing Data Driven Decision Making. NEHAS PUBLICATIONS, Bengaluru (2024)
4. Hawamdeh, S.S.: Cybersecurity and data privacy laws: Balancing innovation and protection in the digital age. Middle East J. Econ. Law Soc. Sci. (MEJELSS), **1**, 36–46 (2025)

Chapter 3
Exploring the Internet of Things (IoT)

3.1 Introduction

Rapidly developing technologically, the Internet of Things (IoT) is a massive network of linked devices capable of data collection and sharing. Broadly, IoT is the ecosystem in which embedded sensors, software, and other technologies link and interact with one another and the cloud via the Internet from actual items. From basic machine-to-many communication to a complex network of smart gadgets influencing many facets of everyday life and business, this idea has grown well beyond its original scope.

IoT originated in the early 1980s and 1990s, connected device implementations, and gained significant pace in the 2000s with developments in wireless technology and the Internet's explosion. Early advances concentrated on automating basic activities and increasing machine efficiency [1]. Technological developments, including IPv6, sensor downsizing, and big data analytics throughout time, have driven IoT into mainstream acceptance across many different fields (Fig. 3.1).

IoT has key traits in connection, intelligence, and interaction. Data interchange is made easier by devices linking with each other, and the Internet is made possible by connectivity. The capacity of tools to gather data, evaluate it, and make judgments depending on predefined rules or learned patterns determines intelligence. Interaction in the context of devices, systems, and people is the smooth integration and communication that generates some interactive and responsive surroundings.

One cannot fully appreciate the value and influence of IoT. By improving efficiency, productivity, and user experience, it has transformed several fields, including healthcare, transportation, agriculture, and smart cities, including IoT, which allows predictive maintenance, real-time monitoring, and automation, thereby saving a lot of money and producing better results. Its influence reaches society at large as it stimulates creativity and changes how people interact with their surroundings

H. Taherdoost, *Innovating and Integrating Emerging Technologies*, Signals and Communication Technology, https://doi.org/10.1007/978-3-031-93953-2_3

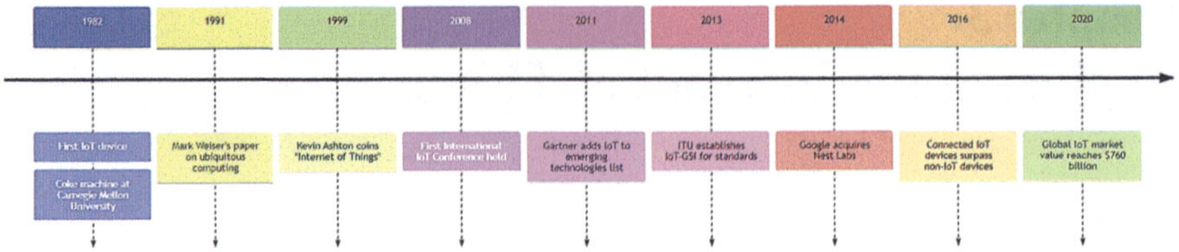

Fig. 3.1 History of IoT evolution

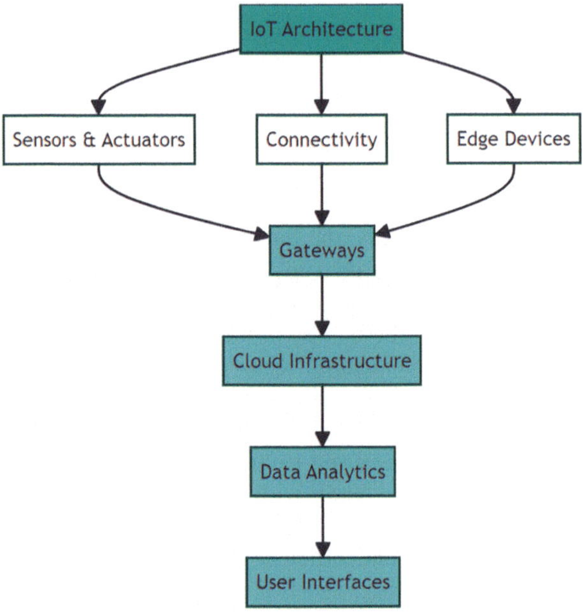

Fig. 3.2 IoT architecture overview

and technology. IoT's ability to propel society forward and economic development remains great and hopeful as it develops.

3.2 IoT Architecture

The Internet of Things (IoT) has numerous basic components whose combined gathering, processing, and acting upon data are achieved (Fig. 3.2). IoT architecture revolves mostly around sensors and actuators. Data from the physical surroundings is gathered by sensors, which also measure motion, temperature, and humidity. Conversely, actuators operate according to processed knowledge, like valve or motor control. Another very vital element is connectivity, which guarantees IoT devices' ability to interact with central systems and one another. This covers many network technologies like Bluetooth, Zigbee, cellular networks, and LPWAN (low-power wide area network) [2]. Edge devices are important as they conduct first data processing near to data generation. This lowers bandwidth use and delay, therefore generally increasing system efficiency. Between IoT devices and the cloud or central servers, gateways aggregate data from many devices and enable safe data transfer, acting as middlemen. Providing scalability and management of vast amounts of data, the cloud infrastructure acts as a central repository for data storage, analysis, and processing. Furthermore, analytics tools and computational resources are available on cloud systems. Using sophisticated approaches like machine learning and artificial intelligence to find patterns, forecast trends, and enable data-driven decision-making, data analytics is the processing and analysis of the data gathered by IoT devices to extract valuable insights. User interfaces are, at last, the platforms or apps by which consumers engage with IoT devices. Their visual, management, and monitoring features help consumers to properly handle and decipher IoT data.

Many times, IoT architecture is explained using layered models to streamline design and deployment. Sensors and actuators at the perception layer, also called the physical layer, collect input from the surroundings and forward it to higher levels for processing. Comprising communication protocols and network technologies guaranteeing safe and effective data transfer, the network layer manages data flow and communication between IoT devices and central systems. Sometimes coupled with the network layer, the edge computing layer consists of locally processing edge devices. This method lowers latency and improves reaction times, therefore reducing the need to transmit vast volumes of data to the cloud. Featuring application software that offers capabilities like data visualization, monitoring, and control, the application layer is where data is evaluated and converted into actionable insights. Focused on the business elements of IoT, including integration with corporate systems, decision-making processes, and value-added service delivery, the business layer sits highest on the hierarchy. This layer makes sure IoT capabilities complement corporate plans and objectives [1].

Different types of IoT communication may be distinguished depending on the way systems and devices share data. Appropriate for situations requiring real-time communication and minimal latency, the device-to-device (D2D) architecture lets devices interact directly with one another without depending on a central server or cloud. Ideal for situations requiring centralized data processing and storage, the device-to-cloud (D2C) architecture consists of devices delivering data straight to the cloud for processing and analysis. Devices provide data to a local gateway in the device-to-gateway (D2G) concept, which then distributes the data to the cloud or another system. This method makes local processing possible and helps to lower data transferred to the cloud's volume. Supporting instances wherein data from many sources has to be consolidated and handled centrally, the gateway-to-cloud (G2C) architecture entails gateways interacting with the cloud to deliver aggregated data from many devices [3].

In Internet of Things systems, interoperability, security, and effective communication depend on several protocols and standards. Often employed in IoT applications for its simplicity and efficiency, MQTT (Message Queuing Telemetry Transport) is a lightweight messaging protocol intended for low-bandwidth, high-latency networks. Designed to run efficiently in contexts with restricted resources and usually utilized in IoT applications, CoAP (Constrained Application Protocol) is suited for limited devices and networks. The Hypertext Transfer Protocol, HTTP/HTTPS, is the standard for online communication; HTTPS offers a safe variant of HTTP for encrypted data transfer. Usually employed in home automation and industrial control systems, Zigbee is a low-power, low-data-rate wireless communication technology used in short-range networks. Common in wearable technology and personal area networks, Bluetooth Low Energy (BLE) is intended for low-energy applications. Appropriate for low data rates and long-range connection, LoRaWAN, Long Range Wide Area Network, is a technology for long-range communication in IoT applications. The most recent iteration of the Internet Protocol, IPv6, offers a large address range to fit the rising count of Internet of Things devices, therefore facilitating direct addressing and communication. Designed for low-rate wireless personal area networks (LR-WPANs), IEEE 802.15.4 is the foundation for protocols such as Zigbee and Thread and finds utility in many IoT applications needing low-power and short-range communication [4].

3.3 IoT Devices and Hardware

IoT systems cannot exist without the field of Internet of Things (IoT) hardware and devices. Sensors and actuators form the foundation of these systems as they are very important for data collecting and physical interaction with the surroundings. Data from the surroundings, such as temperature, humidity, or motion, as well as other measurements, such as those of sensors, are gathered by them and converted into signals fit for processing. Conversely, actuators are employed to produce physical changes depending on commands received from a central system, including opening a valve, changing a motor, or turning on a switch.

Key parts of controlling the connectivity between IoT devices and the larger network architecture are IoT gateways. These devices compile data from several sensors and actuators and forward it to central systems or cloud-based platforms for further processing. Gateways also manage data aggregation and protocol translating so that data from several devices may be easily combined and examined.

The foundation of many IoT devices is embedded systems and microcontrollers. These little computers are used to do certain chores and manage IoT device hardware components. Often integrating sensors, actuators, and communication modules into a single board, embedded systems enable effective IoT function operation and administration. Microcontrollers, microprocessors with built-in memory and input/output peripherals are especially helpful in running predefined tasks and controlling device activities.

Communication between IoT devices and the internet or other devices is enabled by connectivity technologies. Among the many wireless and wired systems, these technologies include Wi-Fi, Bluetooth, Zigbee, and cellular networks. Every

Table 3.1 Key components of IoT devices and hardware

Component	Description	Function	Examples
Sensors	Devices that collect data from the environment by measuring physical properties such as temperature, humidity, and motion	Converts physical measurements into signals for processing	Temperature sensors, motion detectors, humidity sensors
Actuators	Devices that perform physical actions in response to commands from a central system	Executes actions like opening a valve or adjusting a motor	Electric motors, solenoids, servos
IoT gateways	Devices that manage communication between IoT devices and the broader network. They handle data aggregation, protocol translation, and data transmission	Aggregates and transmits data from various sensors and actuators to central systems	Raspberry Pi, industrial gateways
Embedded systems	Compact computing units are designed for specific tasks, integrating sensors, actuators, and communication modules into a single unit	Manages operations and tasks within IoT devices	Arduino boards, custom IoT modules
Microcontrollers	Microprocessors with built-in memory and peripherals that execute specific tasks and control hardware components within IoT devices	Executes predefined tasks and manages device operations	Atmel ATmega, STM32 series
Connectivity technologies	Various protocols enable communication between IoT devices and the internet or other devices. This includes both wireless and wired options	Facilitates data transfer and communication within IoT systems	Wi-Fi, Bluetooth, Zigbee, LTE, Ethernet

networking method is selected depending on range, battery consumption, and data transmission needs; each offers advantages. An important factor in the design and execution of IoT solutions is the choice of connection technology, which affects the general performance and efficiency of IoT systems. Table 3.1 succinctly summarizes the main components of IoT devices and hardware, including their roles, purposes, and examples, therefore offering a clear picture of their contributions to IoT systems.

3.4 IoT Communication Protocols

Within IoT, effective and dependable data flow between devices and systems depends critically on communication protocols. Among the most often used protocols are MQTT, CoAP, HTTP/HTTPS, and many wireless technologies, including Bluetooth, Zigbee, and Z-Wave.

Designed for low-bandwidth, high-latency networks, MQTT, Message Queuing Telemetry Transport, is a lightweight messaging protocol. Operating under a publish-subscribed approach, it lets other clients subscribe to topics to get pertinent updates while clients may post messages to topics. MQTT is perfect for situations needing real-time data transfer and low-power consumption, as it is rather efficient for IoT applications because of its minimum overhead and support of persistent sessions [5].

Another important protocol designed for limited environments, that of Constrained Application Protocol, or CoAP, found in many Internet of Things projects is designed to operate with low-power and low-bandwidth networks; CoAP uses a client-server paradigm more akin to HTTP. Designed for machine-to-machine (M2M) communication, it includes capabilities like asynchronous messaging and resource discovery, qualities that help devices with low processing capability [4, 5].

Simplicity and extensive interoperability with online services make HTTP (Hypertext Transfer Protocol) and its secure variant, HTTPS, quite popular in IoT. The fundamental protocol of the web, HTTP, offers a consistent means of communication for servers and customers via request and response messages. Crucially for safeguarding private data sent by IoT devices, HTTPS, a variation of HTTP with SSL/TLS, guarantees a safe connection over the internet.

Particularly in home automation and sensor networks, Bluetooth, Zigbee, and Z-Wing are wireless communication technologies often used in IoT devices. Bluetooth is fit for devices like smartphones, wearables, and other personal gadgets because of its low-power, short-range communication features. Operating in the 2.4 GHz range and built for low-power, low-data-rate applications, Zigbee provides strong performance in mesh network setups. Another mesh networking system, Z-Wave, runs in sub-1 GHz frequencies and is designed for low-power, dependable connectivity in smart homes.

From basic sensor networks to sophisticated, secure communications across varied and limited contexts, these protocols each offer unique qualities and benefits that fit various kinds of IoT uses (Fig. 3.3).

3.5 IoT Platforms and Software

Operating systems tailored especially for the Internet of Things (IoT) provide the fundamental software layer for IoT device management and control. These operating systems are tailored for IoT contexts, with restricted processor capability, memory, and battery life, among other limitations. Often seen in embedded systems, examples include real-time operating systems (RTOS) such as FreeRTOS and Zephyr, which provide lightweight, effective management of device resources [6].

Offering scalable and flexible resources for storing, processing, and analyzing the enormous volumes of data produced by IoT devices, cloud platforms, and services is vital in the IoT ecosystem. Data storage, analytics, and application development tools abound on these sites, Amazon Web Services (AWS IoT), Microsoft

Fig. 3.3 IoT communication protocols overview

Azure IoT, and Google Cloud IoT. They allow the deployment of apps and services in a scalable, safe environment and easy interaction with many IoT devices.

Emerging concepts called edge and fog computing bring processing and data storage nearer to the source of data production, therefore addressing the constraints of conventional cloud computing. Edge computing improves real-time processing capability by processing data on local devices or gateways at the edge of the network, therefore lowering latency and bandwidth utilization. By building a distributed computing layer between the edge and the cloud, fog computing expands this idea by enabling more sophisticated data processing and analysis nearer where data is produced.

Essential parts of IoT systems, data management, and analytics provide the tools and methods required to manage the enormous volumes of data gathered from Internet of Things sensors. Good data management is data organization, storage, and security that guarantees data availability for analysis. Then, analytics technologies help derive useful insights from this data, enhancing operational effectiveness and decision-making. Machine learning and artificial intelligence, among other advanced analytics, may help IoT systems be more functional, trend-predictive, and process-optimizing tools (Fig. 3.4).

3.6 Security and Privacy in IoT

Given the ubiquitous and linked character of Internet of Things (IoT) devices and systems, security and privacy must be the first priority. IoT settings provide many risks and obstacles that must be resolved if sensitive data is to be kept in integrity and operations to be maintained (Fig. 3.5).

Common security risks and vulnerabilities in IoT are many and may have grave effects. Unauthorized access, data breaches, and cyberattacks, including Distributed Denial of Service (DDoS) assaults, include these hazards. Because of their poor security policies and continuous connection, IoT devices are often targets for attackers. Typical weaknesses include poor or default passwords, unsafe methods of contact, and insufficient software upgrades. Many IoT devices also lack strong security mechanisms, which makes them easily targets for abuse [7].

IoT security depends critically on authentication and authorization. While authorization decides the degree of access given to authorized entities, authentication guarantees that devices and users are who they claim to be. Verifying user and device identification depends on strong authentication systems, including multi-factor authentication (MFA). Authorizing is establishing access limits and rights to guarantee that certain devices or data may only be accessed by authorized organizations. By putting these policies into effect, one guarantees that only trusted devices and people may execute important tasks within the IoT network and help prevent illegal access.

Protecting data sent and kept by Internet of Things devices depends mostly on data encryption and integrity. Encryption guarantees that data is turned into an

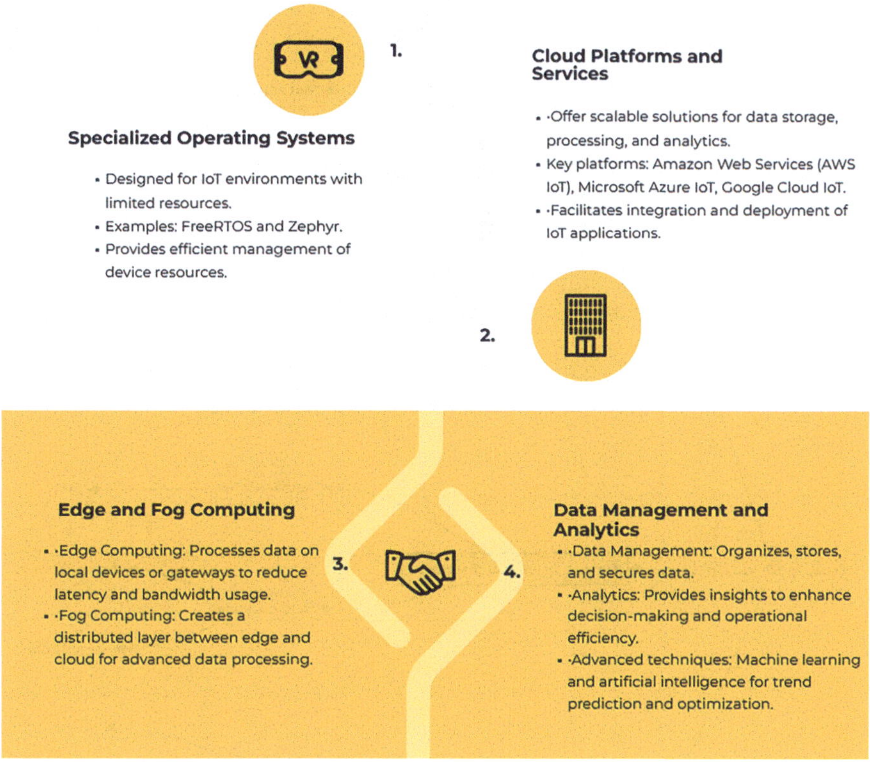

Fig. 3.4 IoT platforms and software overview

unreadable form, therefore making it unavailable to illegal users. This is very vital for protecting private information both on rest and in transit. Data integrity is making sure information is correct and unaffected both during storage and transfer. Data integrity is checked using methods like digital signatures and cryptographic hash, thus guaranteeing that it has not been altered. Strong encryption methods and data integrity checks assist in preventing data modification and eavesdropping.

In IoT, privacy concerns and solutions address the gathering, storage, and usage of sensitive and personal data. The vast amounts of data IoT devices gather, personal information, behavioral patterns, environmental data, raise privacy problems. Implementing privacy-by-design ideas, which include privacy elements in IoT system operation and design, is crucial in order to address these issues. Data anonymizing, that is, removing personal identifiers from data sets, as well as user permission systems guarantee that people are aware and may, therefore, decide how their data is used. Frequent audits and adherence to privacy laws like GDPR help to improve personal data security even further.

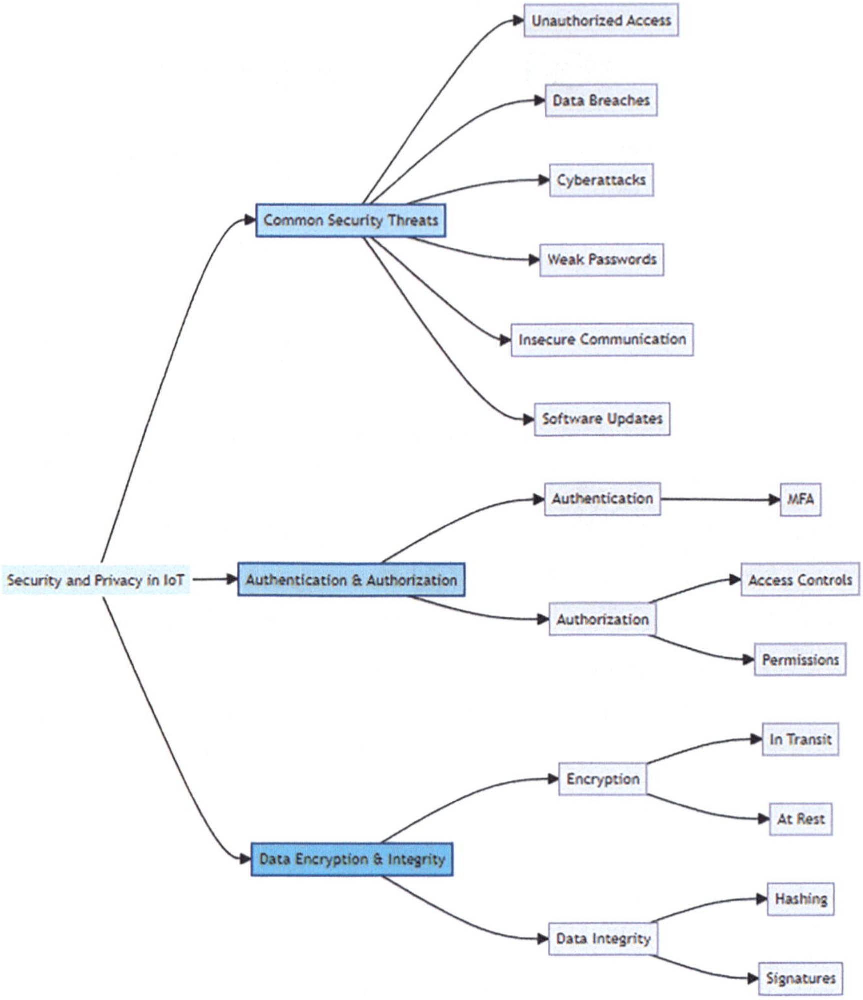

Fig. 3.5 Key components of security and privacy in the Internet of Things (IoT)

3.7 IoT Applications and Use Cases

Applied in many different fields, the Internet of Things (IoT) offers special advantages and revolutionizes sectors. Within the context of smart homes and consumer IoT, technologies are meant to improve domestic convenience, security, and energy economy. Smart thermostats, lighting systems, security cameras, and voice-activated assistants are among examples that let homeowners remotely manage and monitor their surroundings via smartphone applications or voice commands.

Industrial IoT (IIoT) is IoT technology used in industrial environments to increase operating efficiency, safety, and predictive maintenance. To track manufacturing processes, control supply chains, and monitor equipment, solutions often call on sensors and data analytics. Smart sensors, for example, may forecast equipment breakdown before it happens, therefore lowering downtime and maintenance expenses; real-time data analytics maximize manufacturing processes and inventory control.

Medical IoT apps help to improve patient care by means of simplified medical procedures. Wearable health monitors assess vital signs and physical activity, therefore offering useful information for tailored health management. Examples of these monitors include smartwatches and fitness trackers. Remote monitoring tools, among other medical IoT devices, help doctors to follow patient states in real time, therefore enhancing diagnostic accuracy and treatment results. Using IoT, telemedicine technologies help to provide patient management and remote consultations.

Within the framework of smart cities and urban IoT, technologies are used to improve infrastructure management and maximize resource allocation, thereby enhancing urban life. Among them are intelligent street lighting that changes depending on ambient light and activity, waste management systems that track bin levels and maximize collecting paths, and smart traffic control systems that ease congestion and enhance safety. These projects help to provide improved quality of life for the people by supporting more sustainable and efficient metropolitan settings.

IoT applications in environmental monitoring and agriculture center on improving sustainability and best use of resources. Automated irrigation systems and soil moisture sensors, among other smart farming technologies, enable farmers to better control water use and raise crop yields. Using IoT sensors, environmental monitoring systems track air and water quality, weather patterns, and other environmental data, thereby helping conservation initiatives and guaranteeing regulatory compliance.

3.8 Challenges and Future Directions

IoT technology develops under the influence of numerous issues and future directions. Technical difficulties include data security, interoperability, and scalability. Realizing IoT's full potential depends on IoT devices and platforms being seamlessly integrated and communicated upon. IoT devices provide enormous amounts of data that need strong management and analytics tools to glean useful insights. Maintaining device dependability, controlling power consumption, and handling complexity in large-scale installations define operational issues.

As IoT expands, regulatory and ethical issues take the stage. Issues of data privacy, security, and user permission have to be taken care of if we are to safeguard personal data and guarantee moral data usage. While changing laws seek to address these issues, harmonizing rules across many areas and sectors remains difficult. Fostering confidence and acceptance of IoT technology depends on following data protection rules like GDPR and using ethical data practices.

IoT capabilities are being improved by newly developing trends and breakthroughs, including edge computing, artificial intelligence (AI), and blockchain technology. While artificial intelligence helps intelligent decision-making at the edge of a network, edge computing allows real-time data processing. Blockchain presents possible means of protecting IoT transactions and data. Faster and more dependable communication for IoT devices will come from the evolution of 5G technology, hence increasing the range of IoT applications possible.

IoT has a transforming potential for many different fields. Advances are supposed to deliver more customized user experiences, more data analytics, and more automation. IoT, combined with technologies like augmented reality (AR) and virtual reality (VR), will provide fresh creative possibilities. Furthermore, an emphasis on sustainability and energy economy will inspire the creation of environmentally friendly IoT solutions, therefore helping to create a more linked and green planet.

3.9 Developing IoT Solutions

Creating useful IoT solutions calls for a methodical process spanning numerous phases: planning, design, development, testing, implementation, and maintenance. Developing a strategy and defining project goals and criteria, the planning stage and design consists of choosing suitable technology and building a thorough architecture. Hardware and software component assembly and integration take the front stage in development. Testing guarantees the performance as planned and satisfies quality criteria. Deployment is bringing the solution to end users; maintenance guarantees constant support and updates.

Designing IoT devices calls for producing both hardware and software elements. Along with building firmware and user interfaces, this procedure includes choosing sensors, microcontrollers, communication modules, and other parts. Before actual manufacturing, prototyping lets one test and validate design ideas. Iterative testing and improvement guarantee the gadget satisfies performance and functional criteria.

Developing software for the Internet of Things entails designing tools and platforms for interaction with IoT devices. This entails creating firmware for cloud-based apps for data processing, storage, and analysis, as well as for devices. One must guarantee compatibility with several operating systems and communication channels. Developing software also involves designing user interfaces that provide device data and controls easy access.

Implementing and configuring devices in the target environment guarantee appropriate connection and integration with current systems. Monitoring device performance, handling problems, implementing upgrades, and data security management constitute post-deployment maintenance. Maintaining the dependability and functioning of the IoT system all through its lifetime depends on constant support.

3.10 Conclusion

Dynamic and fast-developing, the Internet of Things (IoT) has the ability to change many facets of everyday life and business operations. The in-depth analysis of IoT applications in this chapter shows its major influence in many fields, including smart homes, industrial operations, healthcare, urban management, and agriculture. Through linked devices and data analytics, each application area shows how IoT technologies are boosting efficiency, enhancing quality of life, and allowing better decision-making.

To fully use IoT technology, nevertheless, their spread raises many issues that have to be resolved. Technical obstacles like scalability, interoperability, and data security call for creative solutions to guarantee flawless running and integration. Protecting user privacy and guaranteeing responsible data usage depend on ethical and regulatory issues. More sophisticated and safe IoT systems are made possible by emerging developments such as edge computing, artificial intelligence, and blockchain, which provide fresh chances and answers to these difficulties.

From first planning and design to deployment and continuous maintenance, IoT systems grow across a whole lifetime. Every step is very important for guaranteeing IoT systems' security, dependability, and functionality. The potential of IoT technology to transform sectors and improve daily lives will grow as it develops. Future developments will probably offer more complex applications, more automation, and better sustainability, thereby further incorporating IoT into the fabric of contemporary life.

All things considered, stakeholders in many different fields must grasp the present situation and future possibilities of IoT if they are to properly profit from it. We can fully use IoT to build smarter, more linked, and more efficient settings by tackling obstacles, welcoming ideas, and using a methodical approach to solve the creation of solutions.

References

1. Vermesan, O., Friess, P. (eds.): Internet of Things: Converging Technologies for Smart Environments and Integrated Ecosystems. River publishers, Aalborg (2013)
2. Chilamkurthy, N.S., Pandey, O.J., Ghosh, A., Cenkeramaddi, L.R., Dai, H.N.: Low-power wide-area networks: a broad overview of its different aspects. IEEE Access. **10**, 81926–81959 (2022)
3. Oñate, W., Sanz, R.: Analysis of architectures implemented for IIoT. Heliyon. **9**(1), e12868 (2023)
4. Alhaidari, F.A., Alqahtani, E.J.: Securing communication between fog computing and IoT using constrained application protocol (CoAP): a survey. J. Commun. **15**(1), 14–30 (2020)
5. Khare, A.: Design and Implementation of Enhanced Message Queuing Telemetry Transport Protocol for Internet of Things (Doctoral dissertation, UPES, Dehradun) (2020)
6. Desbiens, F.: Operating systems. In: Building Enterprise IoT Solutions with Eclipse IoT Technologies: An Open Source Approach to Edge Computing, pp. 243–267. Apress, Berkeley, CA (2022)
7. Hallman, R., Bryan, J., Palavicini, G., Divita, J., Romero-Mariona, J.: IoDDoS-the internet of distributed denial of service attacks. In: 2nd International Conference on Internet of Things, Big Data and Security. SCITEPRESS, pp. 47–58 (2017, April)

Chapter 4
Cloud Computing Essentials

4.1 Introduction

A breakthrough technology, cloud computing significantly alters the way that computing resources are supplied and used. Fundamentally, cloud computing offers, over the internet, often referred to as "the cloud computing services" including servers, storage, databases, networking, software, and more. Without physical infrastructure or thorough IT management, this paradigm lets consumers access and control these resources on-demand. Cloud computing abstracts actual hardware into virtual resources dynamically allocated, scaled, and managed by using virtualization technologies.

Still, the idea of cloud computing had origins in the 1960s (Fig. 4.1). Computer scientist John McCarthy suggested during this time that computing might develop as a utility service, much as water or electricity [1]. This visionary concept prepared the ground for the evolution of what today we consider to be cloud computing. The introduction of ARPANET in the late 1960s, the forerunner of the contemporary internet, helped to further enable the sharing of resources and knowledge, hence establishing more foundation for cloud technology [2].

In the development of cloud computing, the 1990s were a pivotal time. The idea of offering services across a network started to take shape as the internet grew. Founded in 1999, companies including Salesforce.com started providing software products online using the Software as a Service (SaaS) approach [3]. Companies like IBM also explored utility computing models around this time, where computer resources were provided on-demand and more like contemporary cloud service offerings [4].

Modern cloud computing first became a mainstream technology in the 2000s. Launching its cloud platform in 2006, Amazon Web Services (AWS) took a revolutionary action providing services including Amazon S3 for storage and Amazon EC2 for computing capability. With scalable and reasonably priced solutions for

H. Taherdoost, *Innovating and Integrating Emerging Technologies*, Signals and Communication Technology, https://doi.org/10.1007/978-3-031-93953-2_4

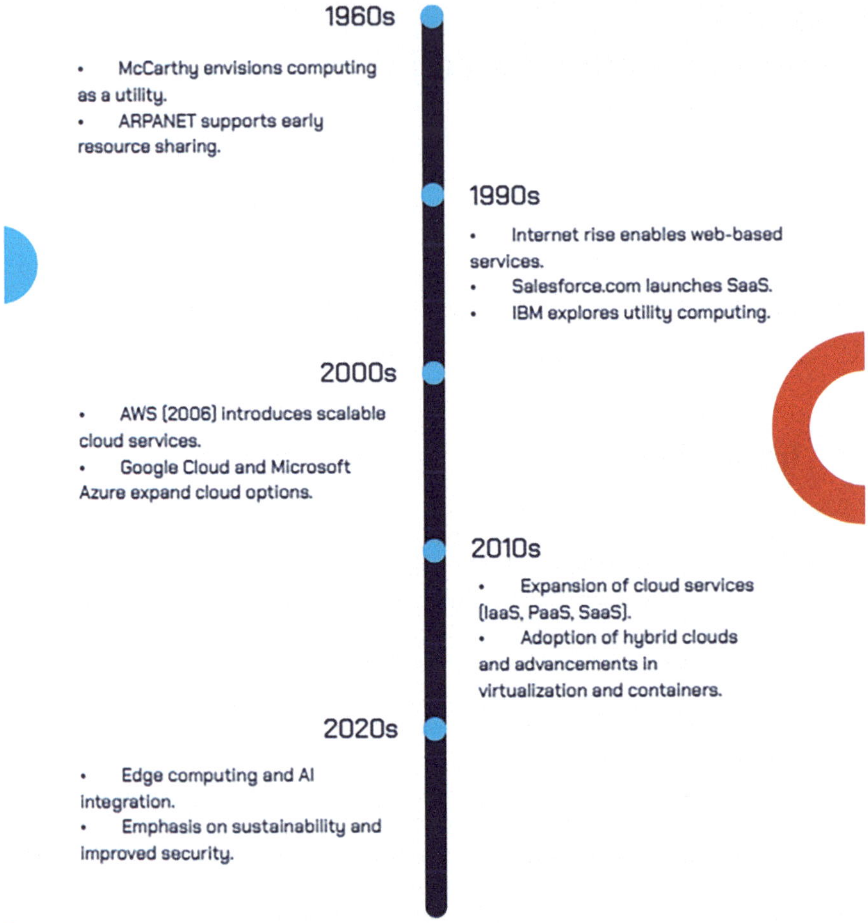

Fig. 4.1 Historical evolution of cloud computing

both companies and individuals, this evolution signaled the start of a new phase in cloud computing. Other big technology companies, including Google and Microsoft, followed AWS and unveiled their cloud platforms, Google Cloud Platform and Microsoft Azure, so augmenting the cloud ecosystem and encouraging competition and invention [4, 5].

Cloud computing grew and embraced quickly as it developed in the 2010s. Cloud service providers presenting a wide spectrum of services and deployment strategies, including Infrastructure as a Service (IaaS), Platform as a Service (PaaS), and SaaS, were proliferating the scene. Combining public and private cloud resources to maximize security, cost, and performance, organizations started implementing hybrid cloud strategies. This era also saw developments in related technologies such as virtualization, containerization (e.g., Docker, Kubernetes), and serverless computing, hence augmenting the capabilities and flexibility of cloud systems [6, 7].

Driven by developing trends and technologies, cloud computing is still changing as we enter the 2020s. Rising popularity is edge computing, which lets data processing near the source cut latency and increase efficiency [8]. More sophisticated data analysis and application development have been enabled by the combination of artificial intelligence (AI) and machine learning services included into cloud platforms. Simultaneously, sustainability is becoming more and more important as initiatives to reduce data center environmental impact and support green cloud technologies take the stage. Continuous developments in cloud security and compliance help to solve the growing issues and legal obligations for data protection and privacy.

4.1.1 Characteristics

Several important traits set cloud computing apart and determine its efficiency and adaptability as shown in Fig. 4.2. On-demand self-service is one of its main features since it lets consumers create computer resources, such as network storage and server time, as needed without involving service providers. This self-service functionality allows resources to be rapidly accessed and changed to fit different applications and workloads.

Still another basic quality is broad network access. From cellphones, tablets, and desktops, cloud services are available via the network from a range of devices that guarantees customers can access to their cloud resources almost anywhere. Standard protocols and technology allowing flawless integration and connectivity help to enable this pervasive access.

The ability of the cloud to serve several consumers utilizing a common pool of resources, such as computing capacity and storage, is predicated on the idea of resource pooling. Demand drives dynamic allocation and reassignment of these resources, hence offering scalability and efficiency. By best using resources and lowering the demand for separate hardware for individual users, this pooling helps to minimize expenses.

A characteristic of cloud services that lets them almost instantly scale resources up or down depending on current demand is fast elasticity. Therefore, although at quieter times, resources can be dialed back to prevent needless expenses, during peak times extra resources can be allocated quickly to meet rising demands. Maintaining performance and managing changing workload demand this dynamic scalability. Measured service is the capacity of the cloud to track and disclose resource consumption, therefore enabling consumers to pay just for what they use. Because users are charged depending on their actual resource consumption rather than a set fee, this metered billing system guarantees cost-efficiency and openness. This quality fits the whole value proposition of the cloud, that of affordability and adaptability.

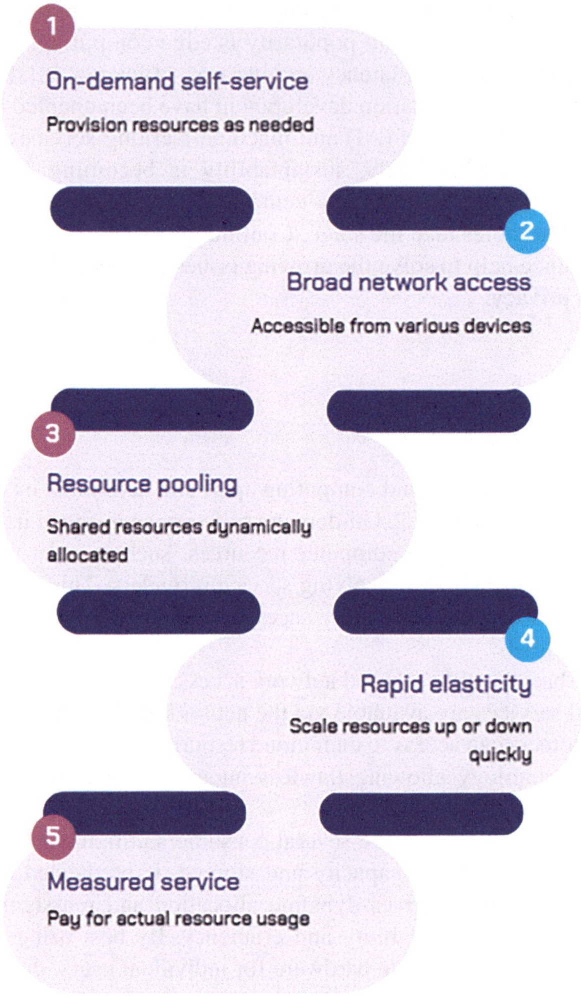

Fig. 4.2 Key characteristics of cloud computing

4.1.2 Service Models

Depending on their demand, cloud computing service models give consumers different degrees of administration, flexibility, and control. Pay-as-you-go, core computing resources including virtual machines, storage, and networks are provided by Infrastructure as a Service (IaaS). Operating systems, apps, and data are within users' control; underlying hardware is not under their control. Companies looking for scalable infrastructure free of the overhead associated with actual hardware administration will find this solution perfect.

Through a development and deployment environment on top of the infrastructure, Platform as a Service (PaaS) offers a higher degree of abstraction. PaaS consumers concentrate on creating applications without regard for the underlying hardware or software layers. Usually including tools for application development, databases, middleware, and other required services, the platform also offers developers who need simplified surroundings to effectively create and implement apps will find this approach appropriate.

Over the internet, Software as a Service (SaaS) provides completely working applications. Users of these programs access them through web browsers without having to install or maintain locally based software. The service provider manages and updates SaaS applications, therefore enabling users to concentrate just on utilizing the program rather than on infrastructure or platform administration. Table 4.1 offers a contrast of the salient traits of cloud service models.

4.2 Cloud Infrastructure

Overcoming geographical constraints, cloud infrastructure lets remote control and robot testing worldwide possible, so improving deployment efficiency by offloading heavy tasks and lowering hardware needs [9]. With an emphasis on managing processing, storage, and networking resources using virtualization techniques, cloud services also include public clouds like AWS, Azure, and private clouds like

Table 4.1 Comparative overview of cloud service models

Feature/characteristic	Infrastructure as a Service (IaaS)	Platform as a Service (PaaS)	Software as a Service (SaaS)
Provision of hardware resources	✓	✗	✗
Provision of software development tools	✗	✓	✗
Management of operating systems	✓	✗	✗
Management of middleware	✗	✓	✗
Management of applications	✗	✗	✓
Management of data	✗	✗	✓
Scalability	✓	✓	✓
Customization	High (e.g., custom virtual machines)	Medium (e.g., limited to provided frameworks)	Low (e.g., predefined applications)
Cost model	Pay-as-you-go for resources used	Pay-as-you-go for platform usage	Subscription-based or pay-per-use
Target users	IT administrators, developers	Developers, project teams	End-users, businesses

OpenStack, therefore providing a range of models from IaaS to SaaS [10]. Green cloud computing presents a way to lessen the environmental impact of cloud infrastructure by encouraging environmentally friendly design, implementation, and management to lower energy consumption and greenhouse gas emissions [11].

4.2.1 Virtualization

A basic concept found in cloud computing, virtualization lets several virtual instances run on a single physical hardware system. Virtualization lets the hardware from the operating system decoupled so that virtual machines (VMs), each running their own OS and applications, may be created. This abstraction layer increases flexibility, lowers cost, and best uses available hardware. Virtualization comes in several flavors: server, storage, and network virtualization. By separating a real server into several virtual machines, server virtualization helps to maximize resource consumption and streamline administration. By pooling physical storage resources, storage virtualization presents them as a single storage object, hence enhancing scalability and efficiency. Abstracting network resources, network virtualization generates virtual networks running free from physical infrastructure, hence improving network flexibility and security.

Among the many advantages virtualization presents are better use of resources, cost reductions, and streamlined administration. It allows fast scalability and effective load balancing by enabling dynamic resource allocation depending on demand. Virtualization does, however, also bring difficulties including possible performance overhead, security issues, and the necessity of certain administration tools. To guarantee best performance and security, efficient implementation calls both rigorous planning and administration. Notwithstanding these difficulties, virtualization is still a vital part of contemporary cloud architecture since it stimulates IT environment efficiency and innovation. Its relevance in cloud computing is highlighted by its capacity to support varying workloads and offer a flexible, scalable, reasonably priced solution.

4.2.2 Storage

Essential part of cloud architecture, cloud storage offers flexible and scalable data storage options. It includes object storage, file storage, and block storage among other forms (Fig. 4.3). File storage is best for shared files and documents; block storage is usually used for databases and applications needing quick, low-latency access. Large volumes of unstructured data, such as backups and multimedia files, fit object storage, the most scalable form.

Cloud storage has many advantages. It provides scalability, which lets companies rapidly increase their storage capacity as needed without making large initial outlay of money. It also offers high availability and redundancy, therefore guaranteeing that data is accessible and hardware failure-protected. Remote access is

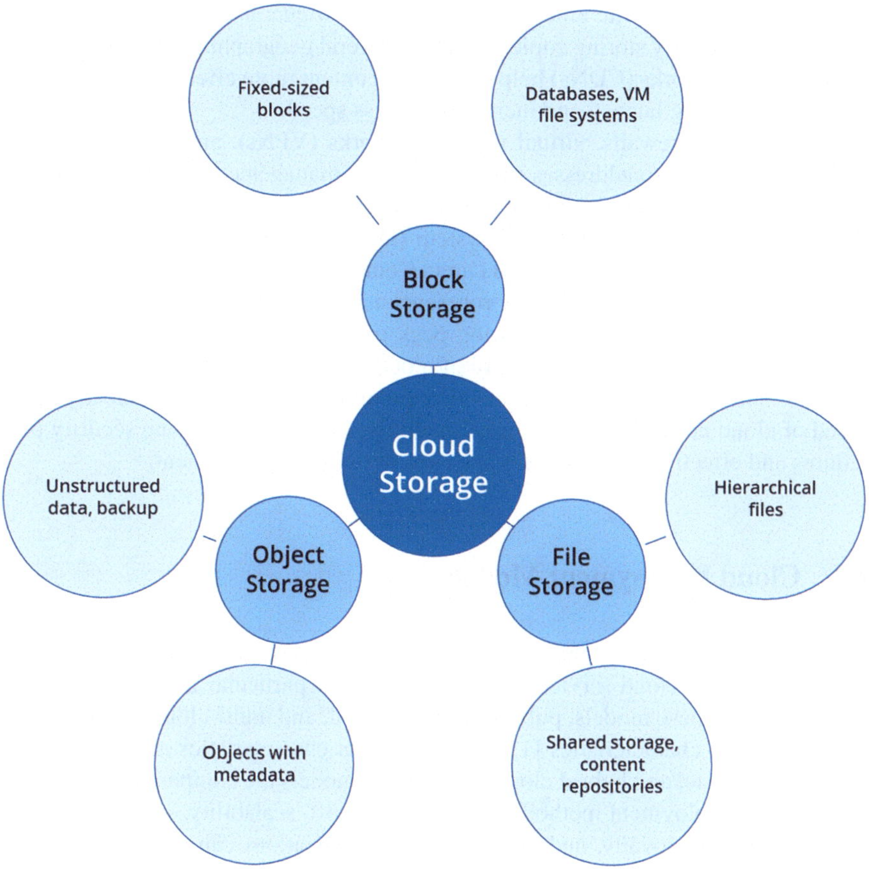

Fig. 4.3 Types of cloud storage

another benefit of cloud storage; users may access their data from anywhere with an internet connection.

Using cloud storage does, however, present several difficulties mostly related to security and conformity. Particularly for sensitive material, data privacy and protection must be first priority. Organizations have to follow different legal rules, which could call for encryption and other security practices. For especially huge datasets, the cost of cloud storage can mount over time and calls for careful planning and administration to maximize expenses.

4.2.3 Networking

Virtual Private Cloud (VPC), which lets users build segregated networks inside the cloud, forms the foundation of cloud computing. This gives over network settings more security and control. Load balancing is also rather important since it helps to distribute incoming traffic among several servers so guaranteeing availability and best performance. By storing copies of data at several geographic locations, Content Distribution Networks (CDNs) help to provide content more effectively to consumers, hence lowering latency and increasing access speed.

By means of firewalls, virtual private networks (VPNs), and security groups, cloud networking also addresses network security management. These instruments guard data and resources from illegal access and online threats. Available from cloud providers are Domain Name System (DNS) administration, which converts domain names into IP addresses, and Direct Connect or ExpressRoute, which enable dedicated links between on-site environments and the cloud.

Whether they are scaling up during peak usage times or scaling down during periods of low activity, the scalability of cloud networking lets companies rapidly adjust to shifting needs. Modern cloud computing infrastructure is mostly composed of cloud networking since its adaptability, together with strong security procedures and effective data transport systems, defines this component.

4.3 Cloud Deployment Models

In cloud computing, cloud deployment models refer to several choices available to companies to use cloud services depending on their particular needs and requirements. Each of these models, public, private, hybrid, and multi-cloud, offers unique advantages and characteristics [12, 13]. The main characteristics and elements of the public, private, and hybrid cloud deployment models are compared in Table 4.2. By selecting deployment methods depending on cost, scalability, security, control, flexibility, vendor loyalty, and other criteria, organizations can guarantee fit with their particular goals and restrictions [13]. Reducing energy consumption and improving sustainability, achieved by means of dynamic workload consolidation and power-aware scheduling [14] depend on optimal resource allocation in cloud-based systems as well. Moreover, strong security measures in cloud installations are crucial; architectures such as Cloud Compute Adoption Architecture (CCAF) offer multi-layered privacy solutions to properly protect corporate data and operations [15].

4.3.1 Public Cloud

Key deployment method in cloud computing, public cloud provides pay-as-you-go services to several users over the internet [16]. For companies and people alike, it is scalable and reasonably priced since it offers several resources like software, infrastructure, and platforms through shared servers [12]. On security, operation, and maintenance, however, novel deployment models such as the proprietary cloud, which combines the benefits of public and private clouds to provide effective

Table 4.2 Comparison of cloud deployment models

Feature/aspect	Public cloud	Private cloud	Hybrid cloud
Definition	Cloud services offered over the public internet and shared among multiple organizations	Cloud infrastructure operated solely for a single organization	A mix of public and private clouds to allow data and applications to be shared between them
Accessibility	Accessible to anyone who subscribes to the service	Accessible only to the organization owning the infrastructure	Combines both public and private access, depending on the configuration
Infrastructure ownership	Owned and operated by third-party cloud providers	Owned and operated by the organization or a third party exclusively for the organization	Public cloud resources are owned by a third party, while private cloud resources are owned by the organization or a dedicated provider
Cost	Generally lower due to shared resources and economies of scale	Generally higher due to dedicated resources and infrastructure	Varies; may be lower than a private cloud but higher than a public cloud depending on usage
Scalability	High scalability due to shared resources	Limited scalability based on owned resources and infrastructure	Flexible; allows scaling in both public and private environments
Security	Security managed by the cloud provider; may have multi-tenant risks	High level of control and customization for security; typically more secure	Offers enhanced security for sensitive data and applications, combined with the scalability of public clouds
Compliance	May meet standard compliance requirements; may not be suitable for highly regulated industries	Easier to meet strict compliance and regulatory requirements	Can balance compliance by keeping sensitive data on private clouds while leveraging public clouds for less sensitive workloads
Management	Managed by the cloud provider	Managed by the organization or a third-party provider	Management requires integration between public and private cloud environments

management and exclusive resources, while lowering maintenance costs, have emerged [17]. Achieving high availability, low latency, and cost-effectiveness in cloud applications depends on an awareness of several deployment paradigms including zonal, regional, multi-regional, global, hybrid, and multi-cloud [18].

4.3.2 Private Cloud

Offering specialized services to one company, private cloud computing gives increased security and data and resource control [17, 19]. It enables flexibility in location and accessibility by letting users use computing resources remotely using server processing capability for running apps and storing data without the need of particular personal PCs [19]. By guaranteeing independent control and exclusive resources for consumers, the private cloud model answers security, operation, and maintenance issues usually observed in public cloud systems [17]. Private clouds provide scalability, flexibility, and cost-efficiency for companies and people by using virtualization, resource pooling, and automation; so, they are a great choice in the scene of cloud computing [12]. Private clouds can also be included into several deployment strategies including proprietary, public, hybrid, and community clouds, thereby offering customized solutions to satisfy different user needs [16].

4.3.3 Hybrid Cloud

Combining elements of public and private clouds, hybrid clouds present a flexible and tailored option for businesses [13, 17]. It lets companies take advantage of both environments, keeping private data on a private cloud and using the scalability and economy of the public cloud for less important processes. This methodology offers a customized strategy that fits certain corporate requirements, therefore addressing issues of security, control, and cost [13]. Popular for companies looking for a balance between security and scalability in their cloud infrastructure, hybrid cloud deployment offers enhanced flexibility, improved data management, and best resource allocation by combining the strengths of public and private clouds [17].

4.4 Cloud Applications

Cloud computing has transformed how companies and people access and make use of computing resources, therefore generating a broad spectrum of applications in many different fields. Through applications like customer relationship management (CRM) systems, enterprise resource planning (ERP) software, and collaboration tools, cloud computing has helped companies in the enterprise domain simplify their operations, lower costs, and improve scalability. These cloud-based solutions help businesses to access strong software and infrastructure without the requirement for large on-site hardware, so enabling remote work and raising general efficiency. Big data analytics, artificial intelligence, and machine learning applications depend on cloud computing now more than ever since it helps companies to glean meaningful insights from enormous volumes of data and make data-driven decisions [20].

Beyond the corporate sphere, cloud computing finds many uses in personal computers, education, healthcare, and scientific inquiry. Common now are cloud storage services, which let people access and save their information from any device with an internet connection. Particularly with relation to remote learning [21], cloud-based learning management systems and virtual classrooms have changed the way teachers and students engage in the classroom. By use of electronic health records

(EHRs) and telemedicine systems, cloud computing has helped the healthcare sector improve patient care and data management [22]. Cloud computing has sped up scientific study by allowing sophisticated simulations, data analysis, and worldwide researcher cooperation [20]. As cloud technology develops, its uses are probably going to get more extensive, inspiring digital transformation and innovation in many other fields.

4.4.1 Enterprise Resource Planning (ERP)

With increased flexibility and access for companies, cloud computing has transformed Enterprise Resource Planning (ERP) systems [23]. With companies looking for ideal designs like service-oriented architecture (SOA) and model view controller (MVC) for effective ERP deployment [24], the integration of cloud-based ERP solutions has become vital in the field of intelligent manufacturing. Just-in-time infrastructure offered by cloud apps helps to provide flawless scalability and workload absorption, therefore simplifying company operations and improving enterprise management [25]. Relative advantage, trust, and IT skills are among the elements impacting the acceptance of cloud-based ERP systems, therefore highlighting the major influence of these technologies on organizational decision-making and operations [26]. The change from conventional on-site ERP to cloud ERP marks a paradigm change in how companies use technology to raise general efficiency and information processing capacity.

4.4.2 Customer Relationship Management (CRM)

Strategic tool used by companies to control relationships with both present and future clients is customer relationship management (CRM). Technology is used to coordinate sales, marketing, customer care, technical support, organization, automation, and synchronization. Among their various benefits is accessibility from any location with internet connection, which allows real-time data sharing and departmental cooperation across multiple entities. Strong analytics and reporting capabilities they also offer enable companies to better understand consumer preferences and behavior, thus enabling more individualized and successful marketing plans [27].

By automating daily chores including email follow-up and appointment scheduling, these technologies free up important time for sales and customer support teams to concentrate on more strategic duties. Though cloud-based CRM offers many advantages, there are certain difficulties including data security issues and the need of consistent internet access. Nonetheless, with appropriate security measures and backup strategies in place, the benefits usually exceed the disadvantages, so cloud-based CRM is a necessary tool for contemporary companies trying to raise client loyalty and pleasure [28].

4.4.3 *Collaboration Tools*

In the modern business, collaboration technologies are indispensable since they let teams operate without problems wherever. These tools cover several uses meant to support team member file sharing, project management, and communication. Tools for cloud-based collaboration offer a centralized platform where staff members may send messages, schedule virtual meetings, and instantly work on documents. These solutions' integration features with other business applications improve their value by enabling simplified processes and higher production.

Accessibility of cloud-based collaboration tools is one of its main advantages; users may access these platforms from any device with an internet connection, so guaranteeing continuous connectivity and the capacity to react fast to problems relating to their employment. These instruments also frequently include strong security measures meant to guard private data and guarantee industry standards' compliance. Notwithstanding these benefits, companies could find it difficult to guarantee customer acceptance, control data privacy, and handle possible technological problems or downtime. Nonetheless, when used properly, cloud-based collaboration tools can greatly improve team cohesion, communication, and general organizational effectiveness, hence they are absolutely essential in the modern workplace.

4.5 Conclusion

Businesses and people now manage and use computing resources in very different ways thanks to cloud computing. Cloud computing increases scalability and flexibility while lowering costs by providing on-demand access to a spectrum of services from virtual machines and storage to sophisticated apps. Its main properties, fast flexibility, and resource pooling help companies to react fast to evolving needs. Knowing the several service models, Infrastructure as a Service (IaaS), Platform as a Service (PaaS), and Software as a Service (SaaS), as well as deployment methods public, private, and hybrid clouds, helps consumers to select the appropriate solution for their particular need. Moreover, ERP, CRM, and teamwork tools among other cloud apps show the useful advantages of cloud usage. Maintaining a competitive edge and fully using cloud computing will depend on being updated about trends and developments as technology develops.

References

1. Dempsey, D., Kelliher, F., Dempsey, D., Kelliher, F.: Cloud computing. In: Industry Trends in Cloud Computing: Alternative Business-to-Business Revenue Models, pp. 9–28 (2018)
2. Rittinghouse, J.W., Ransome, J.F.: Cloud Computing: History and Evolution. Encyclopedia of Information Systems and Technology-Two Volume Set, pp. 178–192. CRC Press, Boca Raton, FL (2015)

3. Ehnholm, S.: The Software as a Service-Business analysis of SaaS model (2009)
4. Marston, S., Li, Z., Bandyopadhyay, S., Zhang, J., Ghalsasi, A.: Cloud computing—the business perspective. Decis. Support Syst. **51**(1), 176–189 (2011)
5. Wyld, D.C.: Moving to the Cloud: An Introduction to Cloud Computing in Government. IBM Center for the Business of Government (2009)
6. Dhayanidhi, G.: Research on IoT threats & implementation of AI/ML to address emerging cybersecurity issues in IoT with cloud computing (2022)
7. Mampage, A., Karunasekera, S., Buyya, R.: A holistic view on resource management in serverless computing environments: taxonomy and future directions. ACM Comput. Surv. (CSUR). **54**(11s), 1–36 (2022)
8. Huh, J.-H., Seo, Y.-S.: Understanding edge computing: engineering evolution with artificial intelligence. IEEE Access. **7**, 164229–164245 (2019)
9. Harinakshi, C., Masooda, A., Suchetha, G.: Cloud infrastructure for robotics: a revolution in robotics development and deployment. In: Shaping the Future of Automation with Cloud-Enhanced Robotics, pp. 20–37. IGI Global, Pennsylvania, PA (2024)
10. Marozzo, F.: Infrastructures for High-Performance Computing: Cloud Infrastructures (2019)
11. Singla, N.: Green cloud infrastructure: mitigating the environmental impact of cloud computing. In: 2023 3rd International Conference on Advancement in Electronics & Communication Engineering (AECE): IEEE, pp. 144–147 (2023)
12. Ghosh, S., Choudhury, R.R., Roy, P., Ghosh, S.: Revealing concepts of a cloud deployment model: a semantic exploration of a new generation of the cloud. In: Emerging Trends in Cloud Computing Analytics, Scalability, and Service Models, pp. 331–339. IGI Global, Pennsylvania, PA (2024)
13. Mishra, S., Awasthi, M., Chandrol, A., Singh, G., Lande, A.: A comparison and analysis of various cloud computing deployment models. In: 2024 IEEE 1st Karachi Section Humanitarian Technology Conference (KHI-HTC): IEEE, pp. 1–7 (2024)
14. Kavitha, J., Rao, P.S., Babu, G.C.: Energy efficient resource utilization of cloud computing environments for deployment models. In: 2023 Second International Conference on Augmented Intelligence and Sustainable Systems (ICAISS): IEEE; pp. 1111–1119 (2023)
15. Yadav, R.K., Natya, S., Mahesh, T.: Cloud deployment unveiled: a security model focused on industry. In: 2023 International Conference on Advances in Computation, Communication and Information Technology (ICAICCIT): IEEE, pp. 923–928 (2023)
16. Tiwari, M.N.: Perspective study of Public Cloud: A highly scalable Cloud deployment model. Authorea Preprints (2023)
17. Zhao, W., Yue, S., Fei, M., Chen, R., Wei, L.: A new cloud computing deployment model: proprietary cloud. In: International Conference on Signal and Information Processing, Networking And Computers, pp. 130–137: Springer, Cham (2022)
18. Berenberg, A., Calder, B.: Deployment archetypes for cloud applications. ACM Comput. Surv. (CSUR). **55**(3), 1–48 (2022)
19. Maaz, M., Ahmed, M.A., Maqsood, M., Soma, S.: Development of service deployment models in private cloud. J. Sci. Res. Technol., **1**, 1–12 (2023)
20. Sunyaev, A., Sunyaev, A.: Cloud computing. In: Internet Computing: Principles of Distributed Systems and Emerging Internet-Based Technologies, pp. 195–236 (2020)
21. Paul, P.K., Ghose, M.K.: Cloud Computing: possibilities, challenges and opportunities with special reference to its emerging need in the academic and working area of Information Science. Proc. Eng. **38**, 2222–2227 (2012)
22. Golightly, L., Chang, V., Xu, Q.A., Gao, X., Liu, B.S.: Adoption of cloud computing as innovation in the organization. Int. J. Eng. Bus. Manag. **14**, 18479790221093992 (2022)
23. Quah, K.W.J., Azhar, Z., Kishan, K.: Cloud enterprise resource planning system and its implication on accountants. In: Eurasia Business and Economics Society Conference, pp. 189–202. Springer, Cham (2022)
24. Lee, H.-Y., Wang, N.-J.: Cloud-based enterprise resource planning with elastic model–view–controller architecture for Internet realization. Comput. Stand. Interf. **64**, 11–23 (2019)

25. Bitkowska, A., Dziembek, D., Gzik, T.: Enterprise resource planning based on cloud computing (Cloud ERP). J. Softw. Syst. Dev. **2024**, 1–16 (2024)
26. Lessa, L., Fekadu, B.D.: Cloud enterprise resource planning (ERP) adoption in Ethiopia: case of an Ethiopian shipping and logistics services enterprise. In: Driving Transformative Change in E-Business Through Applied Intelligence and Emerging Technologies, pp. 165–186. IGI Global, Pennsylvania, PA (2022)
27. Anshari, M., Almunawar, M.N., Lim, S.A., Al-Mudimigh, A.: Customer relationship management and big data enabled: personalization & customization of services. Appl. Comput. Inform. **15**(2), 94–101 (2019)
28. Attaran, M., Woods, J.: Cloud computing technology: improving small business performance using the Internet. J. Small Bus. Entrep. **31**(6), 495–519 (2019)

Chapter 5
Thinking Machines: Artificial Intelligence and Machine Learning

5.1 Introduction

Artificial Intelligence (AI) is a branch of computer science focused on creating systems capable of performing tasks that typically require human intelligence. These tasks include learning, reasoning, problem-solving, perception, and language understanding. AI is not a singular technology but an umbrella term encompassing various techniques and methodologies aimed at mimicking cognitive functions. The ultimate goal of AI is to develop systems that can perform complex tasks autonomously, adapting to new situations and making decisions without human intervention.

The journey of AI began in the mid-twentieth century with the pioneering work of computer scientists like Alan Turing and John McCarthy (Fig. 5.1). Turing proposed the idea of a "universal machine" capable of performing any computational task, which laid the foundation for the concept of a thinking machine. In 1956, McCarthy coined the term "Artificial Intelligence" during the Dartmouth Conference, marking the formal birth of AI as a field of study [1].

Initially, AI research focused on symbolic AI, where programs manipulated symbols to perform logical reasoning and solve problems. Early successes included the development of expert systems and simple AI programs that could play games like chess. However, these systems were limited by their inability to learn and adapt.

The 1980s and 1990s saw the emergence of machine learning (ML), a subset of AI that emphasizes learning from data. ML algorithms enabled computers to recognize patterns and make predictions based on historical data, leading to advancements in areas like speech recognition and image processing. The advent of big data and increased computational power in the 2000s further accelerated ML research, giving rise to deep learning, a powerful technique inspired by the structure of the human brain.

H. Taherdoost, *Innovating and Integrating Emerging Technologies*, Signals and Communication Technology, https://doi.org/10.1007/978-3-031-93953-2_5

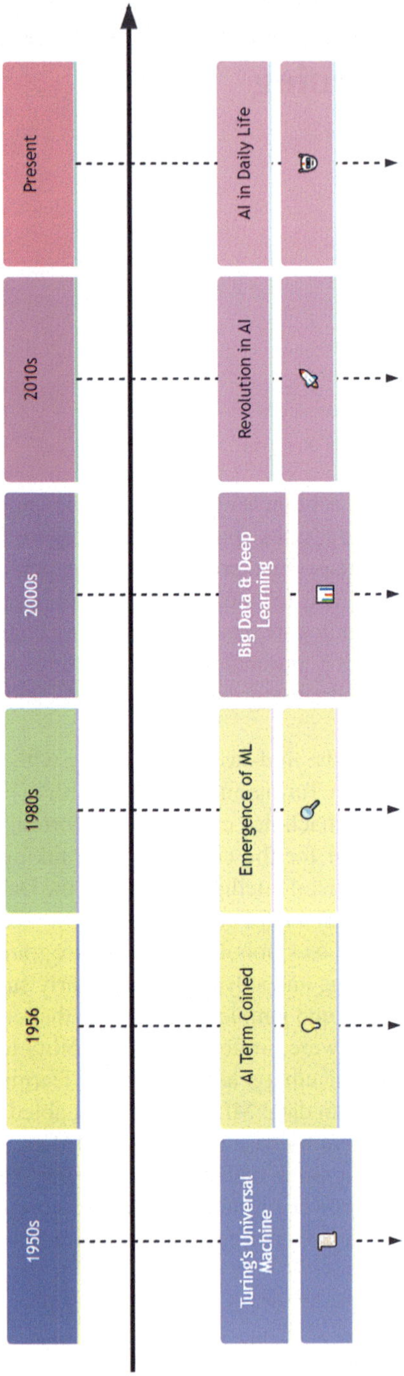

Fig. 5.1 The evolution of AI

Deep learning models, particularly neural networks, have revolutionized AI, enabling breakthroughs in natural language processing, computer vision, and autonomous systems. These advancements have transformed AI from a theoretical concept into a practical technology with widespread applications.

AI has become an integral part of modern society, influencing various aspects of daily life and driving significant technological and economic changes. One of the most notable impacts of AI is its ability to process and analyze vast amounts of data, providing insights that were previously unattainable. This capability is transforming industries such as healthcare, finance, and retail by enabling more accurate diagnostics, personalized recommendations, and efficient operations [1].

In healthcare, AI systems are used to analyze medical images, predict patient outcomes, and assist in drug discovery. For instance, AI algorithms can detect anomalies in radiographs with high accuracy, aiding radiologists in early diagnosis of conditions like cancer. In finance, AI-driven models predict market trends, optimize investment strategies, and detect fraudulent activities, enhancing the stability and security of financial systems.

AI's impact extends to everyday life through applications like virtual assistants, recommendation systems, and smart home devices. Virtual assistants such as Siri, Alexa, and Google Assistant use natural language processing to understand and respond to user queries, making it easier to access information and perform tasks. Recommendation systems employed by platforms like Netflix and Amazon analyze user preferences to suggest content and products, enhancing user experience and driving consumer engagement.

Furthermore, AI is playing a crucial role in addressing global challenges. In agriculture, AI-powered solutions optimize crop yields and reduce resource consumption, contributing to food security. In environmental conservation, AI is used to monitor ecosystems, track wildlife populations, and predict natural disasters, aiding in the protection of biodiversity.

5.2 Core Concepts of AI

Artificial Intelligence (AI) encompasses a broad range of concepts that aim to replicate or simulate human cognitive processes. At its core, AI is concerned with creating systems that can perform tasks requiring human-like intelligence. This includes understanding and generating human language, recognizing patterns, solving complex problems, and making decisions. Unlike human intelligence, which is inherently flexible and context-sensitive, artificial intelligence is limited by its design and the data it processes. AI systems operate based on algorithms and models trained on specific datasets, making their performance highly dependent on the quality and scope of the data they receive.

The Turing Test, proposed by Alan Turing in 1950, is one of the most well-known benchmarks for evaluating a machine's ability to exhibit intelligent behavior indistinguishable from that of a human [2]. In this test, a human evaluator engages

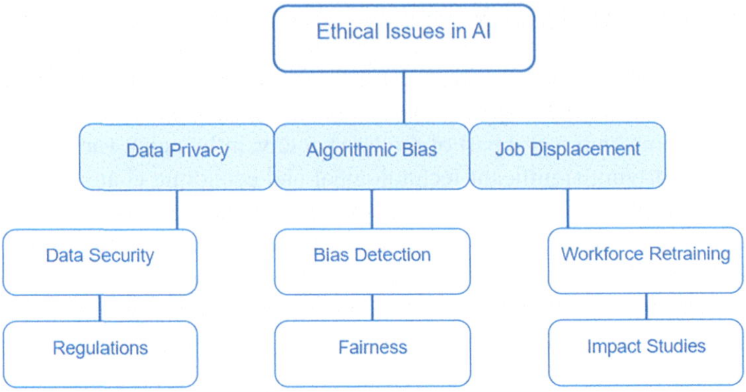

Fig. 5.2 Key ethical issues in artificial intelligence

in a conversation with both a machine and a human, without knowing which is which. If the evaluator cannot reliably distinguish between the two, the machine is said to have passed the Turing Test. This test remains a foundational concept in AI, providing a measure of a system's capacity for human-like interaction. However, passing the Turing Test does not necessarily indicate that a machine possesses general intelligence or understanding; it simply demonstrates that the machine can mimic human responses convincingly in a specific context.

Ethical and philosophical considerations in AI have become increasingly important as the technology advances. Issues such as data privacy, algorithmic bias, and the potential for job displacement due to automation are central to the ongoing debate about the role of AI in society (Fig. 5.2). Ensuring that AI systems are designed and used ethically involves addressing these concerns and developing frameworks that promote fairness, transparency, and accountability. Philosophically, questions about the nature of intelligence, consciousness, and the moral status of machines challenge our understanding of what it means to be intelligent or sentient. As AI continues to evolve, these ethical and philosophical discussions will play a crucial role in shaping the future development and deployment of artificial intelligence technologies.

5.3 Machine Learning Fundamentals

Machine learning is a subset of artificial intelligence that enables systems to learn from data, identify patterns, and make decisions with minimal human intervention. It involves the use of algorithms and statistical models to perform specific tasks by relying on patterns and inference rather than explicit instructions. Machine learning can be categorized into three main types: supervised learning, unsupervised learning, and reinforcement learning (Fig. 5.3).

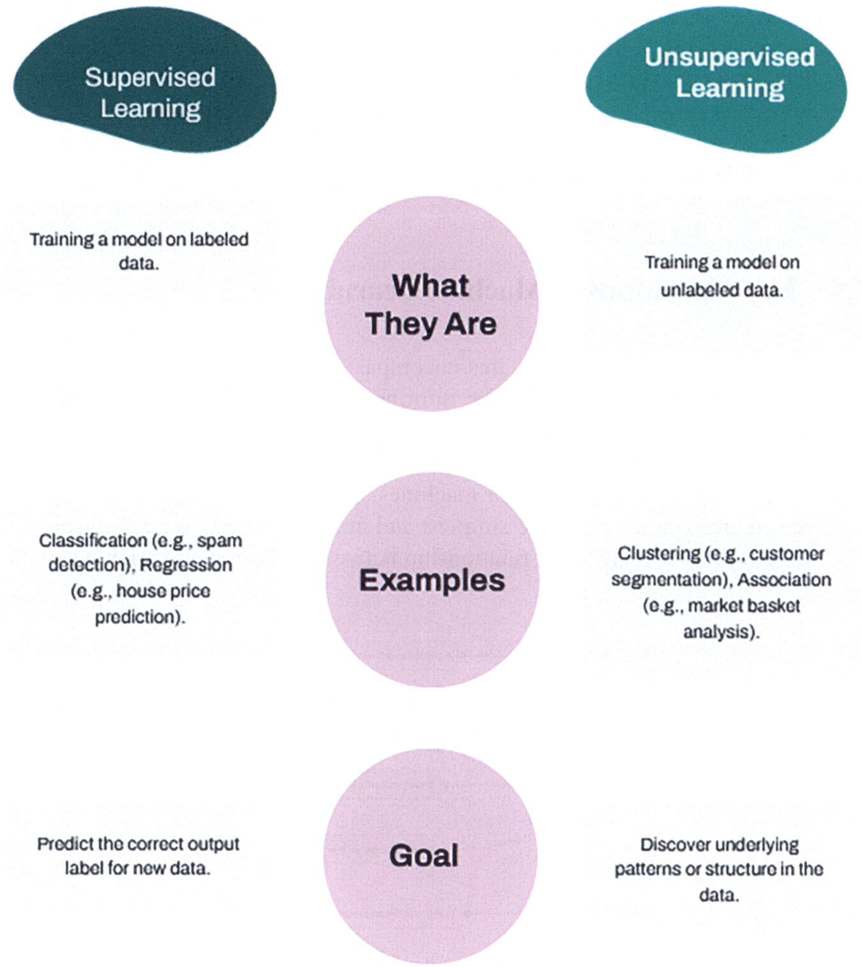

Fig. 5.3 Comparison of supervised and unsupervised learning

In supervised learning, the algorithm is trained on a labeled dataset, meaning that each training example is paired with an output label. The goal is for the algorithm to learn a mapping from inputs to outputs so it can predict the label for new, unseen data. This type of learning is used for tasks such as classification, where the output is a category, and regression, where the output is a continuous value.

Unsupervised learning, on the other hand, deals with unlabeled data. The system tries to learn the patterns and the structure from the input data without any explicit output labels. Clustering and association are two common tasks in unsupervised learning. Clustering involves grouping the data into clusters based on similarities, while association is about discovering rules that describe large portions of the data, such as market basket analysis.

Reinforcement learning is a type of machine learning where an agent learns to make decisions by performing actions in an environment to achieve maximum cumulative reward. The agent receives feedback in the form of rewards or penalties, which it uses to refine its strategy over time. This type of learning is commonly used in fields such as robotics, gaming, and navigation, where the system must make a sequence of decisions to reach a goal.

5.4 Key Algorithms in Machine Learning

Machine learning is a diverse field that encompasses various algorithms designed to analyze data, learn from it, and make informed decisions. Understanding the key algorithms is essential for grasping the fundamental techniques used in machine learning. These algorithms include linear regression, logistic regression, decision trees, random forests, support vector machines, and neural networks (Fig. 5.4).

Linear regression is one of the simplest and most commonly used algorithms in machine learning. It models the relationship between a dependent variable and one

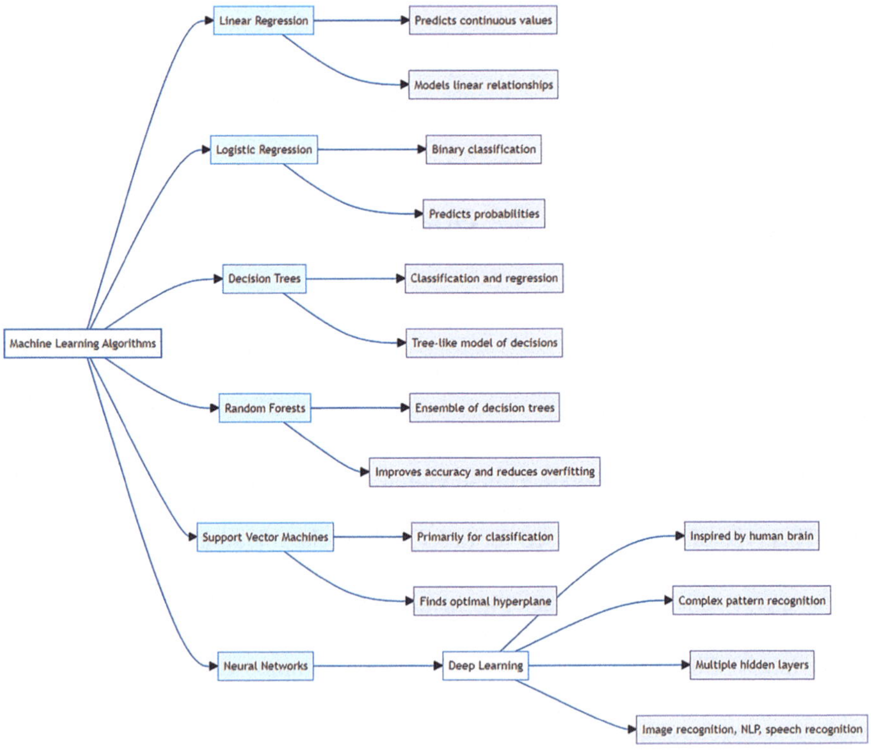

Fig. 5.4 Overview of key machine learning algorithms

or more independent variables by fitting a linear equation to observed data. The primary objective of linear regression is to predict the value of the dependent variable based on the input features. This algorithm is widely used in situations where the relationship between variables is expected to be linear, such as predicting house prices based on size and location.

Logistic regression, despite its name, is used for binary classification problems rather than regression tasks. It estimates the probability that a given input belongs to a particular class. The output of logistic regression is a value between 0 and 1, which can be interpreted as the probability of the input belonging to the positive class. This algorithm is particularly useful for applications like spam detection, where the goal is to classify emails as either spam or not spam.

Decision trees are versatile algorithms that can be used for both classification and regression tasks. They work by recursively splitting the data into subsets based on the values of input features, creating a tree-like model of decisions. Each internal node represents a test on an attribute, each branch represents the outcome of the test, and each leaf node represents a class label or continuous value. Decision trees are easy to interpret and visualize, making them popular for applications where model transparency is important.

Random forests are an extension of decision trees that aim to improve accuracy and reduce overfitting. A random forest is an ensemble of decision trees, typically trained on different subsets of the training data and using random subsets of features for splitting nodes. The final prediction is made by averaging the predictions of individual trees for regression tasks or by majority voting for classification tasks. This approach helps to mitigate the high variance typically associated with single decision trees and results in a more robust model.

Support vector machines (SVMs) are powerful algorithms used primarily for classification tasks, although they can also be adapted for regression. SVMs work by finding the optimal hyperplane that best separates the data into different classes. The algorithm aims to maximize the margin between the hyperplane and the nearest data points from each class, known as support vectors. SVMs are particularly effective in high-dimensional spaces and are widely used in applications such as image classification and bioinformatics [3].

Neural networks, inspired by the structure and function of the human brain, are a class of algorithms designed for complex pattern recognition tasks. They consist of layers of interconnected nodes, or neurons, that process input data and learn to make predictions. Each connection between neurons has an associated weight that is adjusted during training to minimize the prediction error. Deep learning, a subfield of neural networks, involves using multiple hidden layers to capture intricate patterns in data. Deep learning has achieved remarkable success in areas like image recognition, natural language processing, and speech recognition.

The key algorithms in machine learning each have their strengths and specific applications. Linear regression and logistic regression are foundational algorithms for regression and binary classification tasks, respectively. Decision trees and random forests offer interpretable models for both classification and regression. Support vector machines excel in high-dimensional classification tasks, while neural networks and deep learning are indispensable for complex pattern recognition. Understanding these algorithms provides a solid foundation for exploring the vast landscape of machine learning and its applications.

5.5 Data: The Fuel of Machine Learning

Data is the fundamental fuel that powers machine learning. The quality and quantity of data directly influence the performance of machine learning models. Machine learning algorithms learn patterns and make decisions based on the data they are provided. Therefore, having clean, relevant, and abundant data is crucial for building effective models.

In machine learning, data can be categorized into different types, including structured and unstructured data. Structured data is organized and easily searchable, often stored in databases and spreadsheets. This type of data includes numerical values and categorical data, such as customer information, transaction records, and sensor readings. Unstructured data, on the other hand, lacks a predefined format and includes text, images, audio, and video. Examples of unstructured data include social media posts, emails, photos, and audio recordings.

The process of preparing data for machine learning involves several steps, starting with data collection. Data can be collected from various sources such as databases, APIs, sensors, and web scraping. Once collected, data often requires cleaning to handle missing values, remove duplicates, and correct errors. Data cleaning ensures that the dataset is accurate and reliable for training machine learning models.

After cleaning, data needs to be transformed into a suitable format for the algorithm. This step may involve normalization, scaling, encoding categorical variables, and feature engineering. Normalization and scaling adjust the range of numerical data, ensuring that all features contribute equally to the model. Encoding categorical variables converts non-numerical data into numerical form, which algorithms can process. Feature engineering involves creating new features or modifying existing ones to better represent the underlying patterns in the data.

Another critical aspect of data preparation is data splitting. The dataset is typically divided into training, validation, and test sets. The training set is used to train the model, the validation set helps tune the model's hyperparameters, and the test set evaluates the model's performance on unseen data. This division helps ensure that the model generalizes well to new, unseen data and prevents overfitting.

The quality of the data significantly impacts the performance of machine learning models. High-quality data should be representative of the problem domain, diverse enough to cover different scenarios, and free from bias. Data bias can lead to biased models, which in turn can result in unfair or inaccurate predictions. Ensuring data quality involves continuous monitoring and validation to maintain the integrity and reliability of the dataset.

5.6 Applications of AI and ML

Artificial intelligence (AI) has become an integral part of our everyday lives, transforming various sectors and enhancing the way we live and work (Fig. 5.5). In healthcare, AI is revolutionizing diagnostics, personalized medicine, and patient care. Machine learning algorithms analyze medical images to detect diseases with high accuracy, while predictive analytics help in forecasting patient outcomes and optimizing treatment plans. Virtual health assistants and chatbots provide immediate medical advice, improving access to healthcare and reducing the burden on medical professionals.

In the financial sector, AI is used for fraud detection, risk management, and personalized financial services. Machine learning models analyze transaction patterns to identify fraudulent activities in real-time, protecting consumers and financial institutions. AI-driven algorithms assess credit risks and automate trading, enhancing the efficiency and accuracy of financial operations. Personalized financial planning and customer service are also enhanced through AI, providing tailored advice and support based on individual financial behaviors and goals.

The education sector benefits from AI through personalized learning experiences and administrative efficiencies. AI-powered platforms adapt to individual student needs, offering customized learning paths and resources. Automated grading systems save educators time and provide consistent evaluations. AI also supports educational accessibility by offering language translation and speech recognition, aiding students with disabilities and those learning in non-native languages.

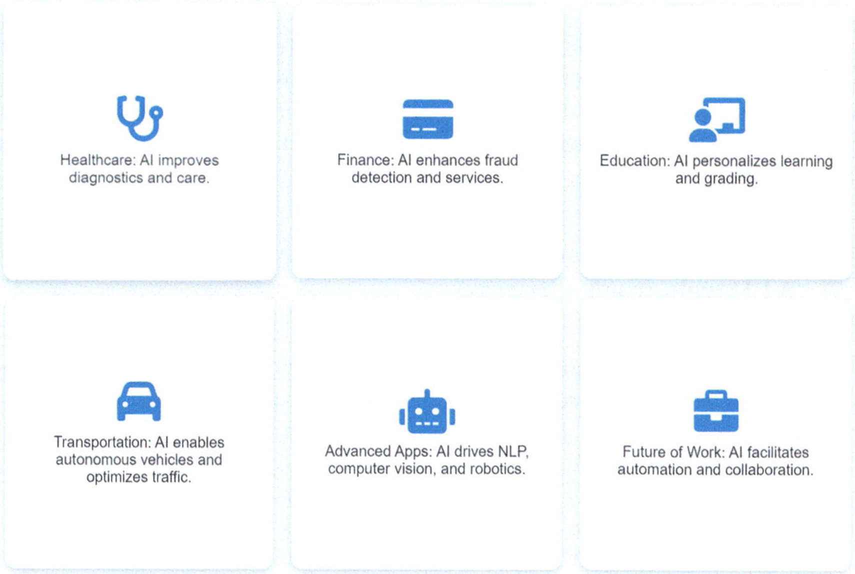

Fig. 5.5 Impact of AI across various sectors

Transportation is experiencing a transformation with AI-driven technologies such as autonomous vehicles and smart traffic management systems. Self-driving cars, powered by advanced computer vision and sensor fusion, promise to reduce accidents and improve traffic flow. AI systems optimize public transportation routes and schedules, enhancing efficiency and reducing congestion. Predictive maintenance algorithms help in identifying potential vehicle failures before they occur, ensuring safer and more reliable transportation services.

Advanced AI applications are pushing the boundaries of what machines can do. Natural language processing (NLP) enables computers to understand and generate human language, powering applications like virtual assistants, chatbots, and language translation services. Computer vision, which involves the interpretation of visual information, is used in facial recognition, image classification, and augmented reality. Robotics and autonomous systems combine AI with mechanical engineering to create machines that can perform complex tasks, from manufacturing to space exploration. In entertainment and gaming, AI enhances user experiences through personalized content recommendations, realistic non-player characters, and dynamic game environments.

AI is also reshaping the future of work, bringing both opportunities and challenges. Automation is transforming jobs by taking over repetitive and mundane tasks, allowing humans to focus on more creative and strategic activities. This shift necessitates the reskilling and upskilling of the workforce to adapt to new roles that require human-AI collaboration. AI is increasingly being used in decision-making processes, providing data-driven insights and predictions that support business strategies and operations. Human-AI collaboration is becoming more prevalent, with AI systems augmenting human capabilities and enhancing productivity in various fields.

As AI continues to evolve, its impact on everyday life and advanced applications will only grow, driving innovation and efficiency across industries. The future of work will be characterized by a symbiotic relationship between humans and AI, where both complement each other's strengths and work together to achieve greater outcomes.

5.7 AI Development Process

The development of artificial intelligence involves several critical steps, starting with defining the problem and gathering the necessary data (Fig. 5.6). This process often begins by identifying a specific issue or opportunity where AI can add value. Once the problem is defined, data collection becomes a crucial step, as the quality and quantity of data directly influence the effectiveness of the AI model. Data can come from various sources, including databases, APIs, and sensors, and must be representative of the problem domain to ensure accurate and reliable results.

After data collection, the next phase is data preprocessing, which involves cleaning and transforming the data to make it suitable for modeling. This includes

Fig. 5.6 Sequential steps in AI development: A visual guide from problem definition to deployment

handling missing values, removing duplicates, and normalizing or scaling numerical data. Data preprocessing ensures that the data is in a format that algorithms can effectively learn from, improving the model's performance and reliability.

Once the data is prepared, selecting the appropriate algorithms and frameworks is essential. Various tools and frameworks are available to support AI development, each offering unique features and capabilities. Popular frameworks such as TensorFlow, PyTorch, and scikit-learn provide libraries and functionalities for building, training, and evaluating machine learning models. These frameworks simplify the implementation of complex algorithms and accelerate the development process [4].

The next step involves training the AI model using the prepared data. During training, the model learns to recognize patterns and make predictions based on the input data. This phase often involves splitting the data into training and validation sets to evaluate the model's performance and make necessary adjustments. Hyperparameter tuning, which involves adjusting parameters to optimize the model's performance, is also a critical aspect of this stage.

After training, testing and evaluating the AI model are crucial to ensure its accuracy and effectiveness. This involves assessing the model's performance on a separate test set that was not used during training. Evaluation metrics such as accuracy, precision, recall, and F1 score are used to measure the model's performance and identify any areas for improvement. Testing helps verify that the model generalizes well to new, unseen data and meets the desired performance criteria.

Deployment and integration follow successful testing. This phase involves deploying the AI model into a production environment where it can be used to make real-world predictions or decisions. Integration with existing systems and workflows is essential to ensure that the AI solution operates seamlessly and delivers the expected value. Ongoing monitoring and maintenance are also necessary to address any issues that arise and to ensure the model remains accurate and effective over time.

5.8 Challenges in AI Implementation

Implementing artificial intelligence (AI) presents a variety of technical challenges that need to be addressed to ensure successful deployment. One of the primary issues is the complexity of developing algorithms that can effectively process and interpret data. This involves designing sophisticated models that can handle large volumes of data and adapt to new, unseen data. Additionally, ensuring the accuracy and reliability of these models requires rigorous testing and validation processes. Computational resources also play a critical role, as training advanced AI models often demands significant processing power and memory, which can be both costly and resource-intensive.

Ethical and social challenges are another significant aspect of AI implementation. As AI systems become more integrated into daily life, concerns about privacy, bias, and fairness grow. For instance, algorithms can unintentionally perpetuate biases present in training data, leading to unfair or discriminatory outcomes. Ensuring transparency in AI decision-making processes and safeguarding user privacy are critical issues that need to be addressed to build trust and accountability in AI systems. Furthermore, there is the challenge of balancing the benefits of AI with its potential impacts on employment and societal structures.

Regulatory and compliance issues add another layer of complexity to AI implementation. Different regions have varying regulations concerning data protection, AI usage, and ethical standards. Navigating this regulatory landscape requires careful attention to ensure compliance with legal requirements and industry standards.

Companies must stay updated with evolving regulations and may need to adapt their AI systems to meet these legal obligations. This involves not only understanding the regulations but also implementing practices that align with legal and ethical guidelines.

5.9 Future of AI and ML

The future of artificial intelligence (AI) and machine learning (ML) is poised to be transformative, with several emerging technologies shaping the landscape. Innovations in AI are accelerating the development of more sophisticated models that can handle increasingly complex tasks. Advances in areas such as natural language processing and computer vision are enabling more intuitive and human-like interactions with technology. Additionally, the integration of AI with other cutting-edge technologies, such as quantum computing and edge computing, promises to enhance computational power and efficiency, pushing the boundaries of what AI can achieve.

Future research in AI and ML is likely to focus on several key areas. One significant direction is the pursuit of general artificial intelligence (AGI), which aims to develop systems with broad cognitive abilities similar to those of humans. Unlike narrow AI, which is designed for specific tasks, AGI would have the capacity to perform a wide range of activities and learn from diverse experiences. Researchers are also exploring ways to improve the interpretability and explainability of AI models, making it easier to understand how these systems arrive at their decisions. Ensuring the robustness and reliability of AI systems under varying conditions remains a critical area of investigation.

The role of AI in society is also evolving, with governance and policy becoming increasingly important. As AI systems become more embedded in daily life, establishing frameworks for responsible use is crucial. This includes developing policies that address ethical considerations, such as privacy, security, and the potential for misuse. Efforts to address bias and fairness in AI are gaining prominence, with researchers and policymakers working to ensure that AI systems do not perpetuate or exacerbate existing inequalities. Ensuring that AI technologies are used for social good involves creating strategies that harness their capabilities to address societal challenges, such as improving healthcare, education, and environmental sustainability.

5.10 Conclusion

As we look back on the journey through the world of artificial intelligence (AI) and machine learning (ML), it is clear that significant strides have been made. We have explored the foundational concepts that drive these technologies, from the basic principles of supervised and unsupervised learning to the more advanced techniques of deep learning and neural networks. Understanding these key concepts provides a solid foundation for appreciating how AI systems are developed and deployed. It also highlights the transformative potential of AI in various sectors, including healthcare, finance, and transportation.

Looking ahead, the future of AI is incredibly promising. The continuous advancement in algorithms, increased computational power, and the proliferation of big data are driving innovations that could reshape industries and enhance human capabilities. AI technologies are expected to become more sophisticated, capable of handling complex tasks with greater efficiency and accuracy. The integration of AI with other emerging technologies, such as the Internet of Things (IoT) and quantum computing, will further amplify its impact, offering new opportunities for solving complex problems and creating value.

Reflecting on the developments in AI and ML, it is essential to remain mindful of the broader implications of these technologies. While the benefits are substantial, the challenges, such as ethical considerations, data privacy, and the need for robust governance frameworks, must be addressed to ensure that AI technologies are developed and used responsibly. By maintaining a balanced perspective and fostering interdisciplinary collaboration, we can navigate the evolving landscape of AI and leverage its potential to drive positive change.

The road ahead for AI and ML is filled with opportunities and challenges. As we advance, it is crucial to stay informed about technological developments, ethical considerations, and societal impacts. By doing so, we can contribute to shaping a future where AI not only enhances technological capabilities but also aligns with our values and goals for a better world.

References

1. Rojas, R.V.B.: Artificial intelligence: genesis, development, and future. In: Revolutionizing Communication, pp. 1–15. CRC Press, Boca Raton, FL (2024)
2. Hernández-Orallo, J.: Twenty years beyond the Turing test: moving beyond the human judges too. Mind. Mach. **30**(4), 533–562 (2020)
3. Brereton, R.G., Lloyd, G.R.: Support vector machines for classification and regression. Analyst. **135**(2), 230–267 (2010)
4. Wang, Z., Liu, K., Li, J., Zhu, Y., Zhang, Y.: Various frameworks and libraries of machine learning and deep learning: a survey. Arch. Comput. Methods Eng., **31**, 1–24 (2019)

Chapter 6
Virtual Reality Technology Advancements

6.1 Introduction

There have been numerous attempts throughout history to create virtual reality (VR) experiences. Together, educators and artists used panoramic paintings that filled their students' field of vision in the eighteenth century to provide the impression of presence. Before the first VR helmet mounted display (HMD) was created in 1960 as a way to show digital sensory content and manage what users see, hear, and feel, there were iterative advances. Over the past 10 years, VR has become increasingly popular. This trend has accelerated since 2016, when a number of businesses started to manufacture affordable and user-friendly gear, making the technology more widely available [1].

From a futuristic idea, VR technology has developed quickly into a common tool used in many different sectors. Fundamentally, VR is a virtual experience that may either replicate the real world or build whole fictitious worlds. Unlike conventional user interfaces, VR puts the user into an experience that allows interaction with 3D worlds in either real world or beyond manner. Hardware and software combining head-mounted displays (HMDs), motion sensors, and computer-generated surroundings help one to attain this immersion. VR's ultimate aim is to produce a sensation of presence whereby users feel as though they are really within the VR.

Over time, VR technology has advanced quickly and is now a vital tool in many different sectors. VR, together with multimedia and network technology, has emerged as one of the most promising computer technologies since its inception less than a century ago. Multidisciplinary domains like computer graphics, perceptual psychology, human physiology, and others have impacted the development of VR, helping to create immersive experiences and perceptual illusions [2]. VR applications have proliferated in fields like scientific research, architecture, and health, providing creative and educational means of presenting data [3]. By mimicking social contexts to improve social functioning and involvement, VR has

H. Taherdoost, *Innovating and Integrating Emerging Technologies*, Signals and Communication Technology, https://doi.org/10.1007/978-3-031-93953-2_6

demonstrated potential in helping people with mental health and substance addiction disorders [4]. The goal of VR technology's ongoing development is to make it more useful, intelligent, and sophisticated so that it may eventually replace physical hardware with virtual hardware.

The power of VR to improve empathy, creativity, communication, and engagement makes it a valuable tool in today's environment [5]. Because it provides immersive experiences and enhances training procedures, it is essential in a number of industries, including business, healthcare, education, and entertainment. VR technology serves as a bridge between the physical and digital worlds, giving users access to virtual environments in which they can interact with and manipulate their surroundings. This has been shown to increase productivity and quality across a range of industries. VR is an essential component of future breakthroughs and discoveries in many sectors since its development has made it possible for an experience that is more participatory and engaging [6, 7].

Though its acceptance is rising, VR technology is still in its early years and presents many possibilities but also problems. Realizing more realistic and immersive virtual environments is one of the main difficulties. Graphics rendering, tactile feedback, and artificial intelligence all have to be much advanced here. Improving the user experience by tackling problems including motion sickness and the physical pain connected with using HMDs for long periods is even another difficulty. Moreover, topics of continuous research and discussion include the social consequences of VR, including its effect on mental health and human interaction.

Future VR technology looks to be bright. The possible uses of VR are projected to grow greatly as hardware gets more powerful and reasonably priced as well as software development tools get more complex. Emerging technologies like augmented reality (AR) and mixed reality (MR) are also probably going to intersect with VR to produce fresh hybrid experiences that mix the physical and virtual worlds. Furthermore, developments in artificial intelligence and machine learning should improve the interactivity and personalization of VR experiences, hence increasing their engagement and fit for individual consumers.

6.2 VR Hardware and Software: Types and Specifications

Any immersive VR experience is built upon VR hardware, which gives users the means to interact with and view a virtual environment (Fig. 6.1). The Head-Mounted Display (HMD), the visual and aural portal to the virtual world, is the most important element. High-resolution screens and lenses built in HMDs provide a stereoscopic 3D effect, so optimizing a broad field of view (FOV) and refresh rates that reduce motion sickness and latency. Apart from visual immersion, some HMDs incorporate integrated headphones for spatial audio, hence boosting the VR's sense of presence.

Motion controllers—which let users interact with objects and negotiate the virtual environment—are another crucial part of VR hardware. Usually using sensors

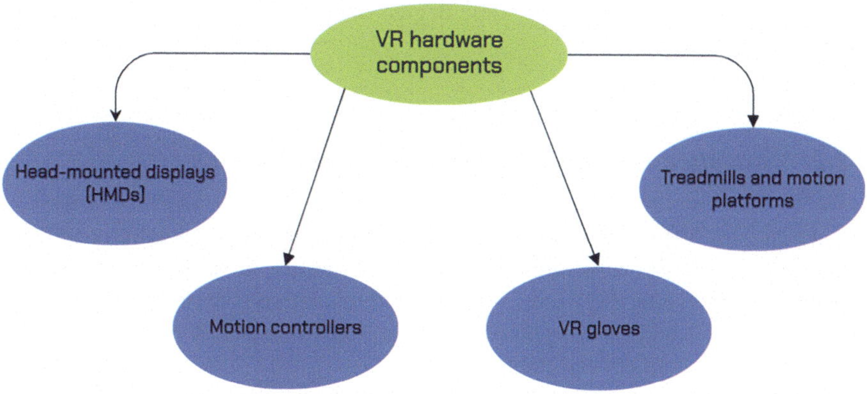

Fig. 6.1 Components of VR hardware

tracking user hand motions and gestures, these controllers convert them into matching actions in the VR environment. More complex systems let users "feel" virtual items by including VR gloves with more exact tracking and haptic feedback. Complementing these are motion platforms and treadmills, which replicate actual movement and enable more natural and immersive experiences.

6.2.1 Types of VR Systems

Three key categories can help one generally classify VR systems: tethered VR, standalone VR, and mobile VR. The most potent, tethered VR systems provide immersive experiences and upscale visuals. These systems are linked to a PC or gaming console, which provide the computing capability needed to run sophisticated VR programs. Popular instances are the HTC Vive and the Oculus Rift. The tethered character of VR worlds lets more realism and complexity possible, but it also limits mobility because of the physical link to the computer.

Between performance and convenience, standalone VR systems like the Oculus Quest line provide a compromise. Built right within the helmet itself, these systems feature all the technology required—namely CPUs, sensors, screens, etc. This removes the requirement for outside devices so users may travel unhindered by a computer. Although standalone VR systems might not equal tethered systems' graphical fidelity, they provide a more user-friendly and accessible experience, therefore making VR more portable and simpler to set up.

Using smartphones as the main gear, mobile VR systems place the phone—like the Samsung Gear VR or Google Cardboard—into a helmet. These systems have low graphics and computing capability even if they are the most reasonably priced and easily available. For casual users or those brand-new to VR, Mobile VR offers a simple introduction to VR experiences. But because of mobile hardware's limits, these devices cannot provide the same degree of immersion as tethered or standalone VR systems.

6.2.2 *Specifications and Performance Benchmarks*

When assessing VR systems' user experience and efficiency, some criteria and performance standards become quite important. Among the most important ones are latency, field of vision (FOV), refresh rate, and resolution (Fig. 6.2). Every one of these components adds to the general quality of the VR experience, therefore influencing the user's comfort and immersion in VR.

Usually expressed in pixels per eye, resolution is the clarity of the pictures seen via the VR helmet. Better image quality directly relates to higher resolution, hence lowering the "screen door effect" whereby people view a grid of lines between pixels. Most modern VR headsets provide resolutions ranging from 1080 x 1200 pixels per eye to 2160 x 2160 pixels per eye or above. Higher resolutions are still driven

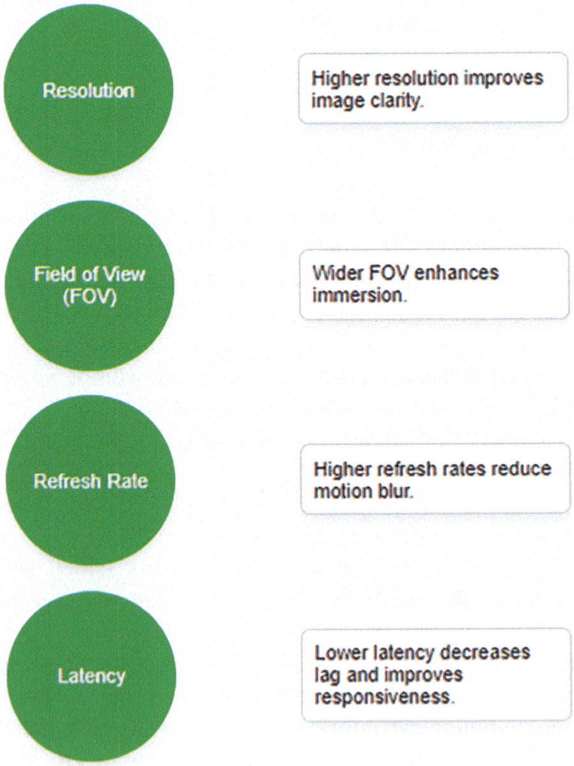

Fig. 6.2 VR hardware specifications

by developers trying to create hyper-realistic landscapes capable of supporting sophisticated simulations and minute visual elements. Rising resolution, however, requires more strong processing capability to produce images that flow naturally, hence balancing performance with clarity.

Another important criterion is Field of View (FOV), which shows the observable world's size at any one point. A larger FOV lets users see more of the virtual environment without rotating their heads, therefore enhancing their immersive experience. Though some high-end versions reach above 120 degrees, standard VR headsets have FOVs ranging from 90° to 110°. Although a larger FOV improves immersion, it can also tax processing resources since showing more of the surroundings in great detail calls for large computing capability.

Measuring in Hertz (Hz), refresh rate is the frequency at which the display changes the image. Higher refresh rates—such as 90 Hz or 120 Hz—ensure smoother movements and help to lower the common VR motion sickness risk. Usually around 60 Hz, the aim is to either match or surpass the natural motion of the human eye. Though they are essential for a flawless and enjoyable user experience, faster refresh rates call for higher processing capability.

Maintaining immersion and comfort ultimately depends critically on latency, the delay between a user's motion and the matching update in the VR world. To avoid obvious lag, which might disturb the experience and create discomfort, latency should preferably be less than 20 ms. Reducing latency calls for hardware and software component optimization including faster sensors, higher data transmission rates, and better rendering approaches.

6.3 Applications of VR in Various Industries

VR has transformed industries by offering immersive experiences in gaming, education, healthcare, entertainment, real estate, and engineering as shown in Fig. 6.3.

6.3.1 *Gaming*

By offering immersive experiences and improving user involvement, VR technology has profoundly affected the gaming sector [8–10]. VR applications in gaming enhance game design, user experience, and technology innovation by allowing players to engage realistically and immersively with game characters and scenes. Different gaming wearable devices have made use of VR technology to let users interact with games using their head, hands, feet, even their whole body, therefore improving the whole gaming experience. Moreover, by including holographic displays and improving film and television special effects, VR in gaming has the power to transform the sector and give consumers an engaging and dynamic gaming experience.

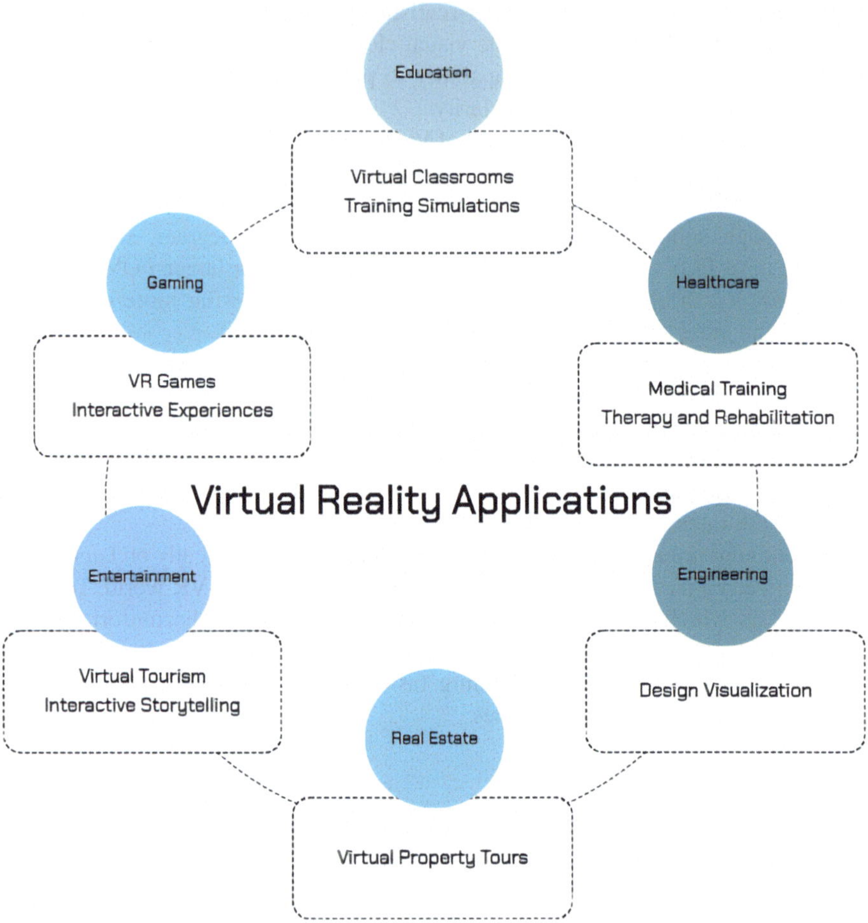

Fig. 6.3 VR across industries

Adoption of VR gaming, meanwhile, offers special difficulties as well. Many users still find great difficulty with motion sickness, usually brought on by differences between real and virtual motion. Developers are always striving to maximize VR experiences so that they lower discomfort and increase accessibility. Some customers may find the great cost of VR hardware to be a deterrent, even if as technology develops prices are progressively declining. Notwithstanding these obstacles, VR gaming has bright future prospects since ongoing innovation is predicted to improve the quality and availability of VR experiences.

6.3.2 Education

VR technology is applied more and more in education to improve learning opportunities. VR enables immersive settings where users may engage and learn from many angles by mixing 3D objects with audio and visual cues [11]. By means of psychological reconstruction of knowledge structures, this technology enables the modification of learning opportunities, therefore benefiting teachers and students alike [12]. Emphasizing the need for involving and creative teaching approaches that VR helps, VR applications in education cover several spheres including educational games, art education, and ideological and political education [13]. By building virtual worlds where students may explore, interact, and learn in ways that traditional approaches cannot mimic, VR in education seeks to make learning more interesting and successful.

VR's capacity to create a virtual environment where students may explore, interact, and engage with materials in previously unthinkable ways helps one of its main advantages in education: VR can, for instance, send pupils to historical landmarks, outer space, or inside the human body, therefore offering a hands-on learning opportunity that improves knowledge and retention. In disciplines like biology, chemistry, and history, where textbooks alone cannot clearly picture, this kind of experiential learning is especially helpful.

Real-world situations are likewise being replicated for teaching via VR. In disciplines like medical training, engineering, and vocational education, where practical experience is absolutely vital, this is especially important. Medical students can, for example, practice procedures in a risk-free virtual environment, which helps them to build confidence and hone their abilities prior to treating real patients. In a similar vein, engineering students can investigate difficult machinery or infrastructure projects virtually, allowing them to grasp minute intricacies and procedures free from the requirement for actual models. By offering useful, real-world applications, this capacity to replicate events helps not only the learning process but also the quality of instruction.

Remote learning made possible by VR is unlike that which conventional online education cannot provide. Students may attend lectures, engage in group discussions, and work on assignments from anywhere in the globe by means of virtual classrooms. By allowing students anywhere to access top-notch learning opportunities, this could help to democratize knowledge. VR is enabling a more inclusive and fair learning environment where students may flourish wherever by removing physical obstacles to education, therefore empowering them.

6.3.3 Healthcare

Through creative technologies for medical teaching, surgery, and patient rehabilitation, VR is revolutionizing healthcare. This technology first emerged in the medical profession in the 1990s to offer sufficient training and ensure the success of the surgical operation. To produce innovation in the medical area, doctor, surgeon, scientists, researcher, and student need greater cooperation and communication. One important tool that links these groups to handle difficult situations is VR. We must find out how it contributes to correct training and prepares a better solution for a medical company. In the medical profession, it appears to be a valuable training tool. Need to explore how students might assist to deliver appropriate therapy to the cancer patient by using the digital experience of complicated surgery in an operating room [14].

Applied in many different ailments, VR is a new technology extensively used in the medical sector in recent years. Actually, this technology represents the simulation of the peripheral environment via a computer as well as communication via a receiver. VR is comprised of output tools (vision, hearing, tactile, and power transmitter), input devices (mouse, chaser, gloves, etc.), a graphical manufacturing system of the VR, and an information program. Based on the type of therapy and individual capacity, all the elements of activity, including length, intensity, and style of feedback, can be adopted in a virtual environment. People also can view their motor outcomes and fix them if needed [15].

6.3.4 Entertainment

By means of immersive experiences that carry users into fascinating, interactive worlds, VR has transformed the entertainment sector. Unlike conventional media, VR lets consumers actively participate in their entertainment, therefore providing a degree of involvement not before possible. In VR gaming, users can interact with game components in a way that replicates real-life activities, investigate more realistic worlds, and engage in intricate narratives. VR is advancing movies and interactive narrative significantly outside of games. Watching scenes from several angles and interacting with characters in real time, VR movies and experiences let spectators enter the tale. This fresh kind of narrative improves emotional involvement and generates a more customized viewing experience.

VR has opened virtual tourism's and theme parks' opportunities. From the comfort of their homes, users can travel far-off places, tour historical sites, or enjoy exhilarating rides. Attractions offer a window into how VR may generate interesting and easily available leisure pursuits. VR technology is probably going to become more important in entertainment as it develops since it provides ever more complex and immersive experiences that challenge our play, viewing, and media interaction capacity.

6.3.5 *Other Industries*

VR is transforming experiences of property viewing in the real estate industry. Without physically visiting homes and businesses, potential purchasers and renters can now access immersive virtual tours of both. By letting customers investigate properties in great depth from any place, this not only simplifies the property search process but also enables them to make more educated judgments. VR presents a great tool for real estate brokers to more successfully present homes and draw more business.

VR is becoming quite helpful in engineering and design. Using VR, architects and engineers may more realistically and intuitively see and interact with 3D models of their projects. More efficient design evaluations and changes made possible by this capacity help to lower expensive mistakes and improve project results before actual building starts. VR also helps teams to work together in a shared virtual environment independent of their real locations, therefore facilitating cooperative design processes.

Advancing in the manufacturing sector is the VR sector. Using VR simulations lets producers digitally prototype and test goods, therefore speeding up the development process and improving the design accuracy. Early in the development cycle, this method helps find any problems, thereby saving money and raising the quality of the products.

6.4 Ethical and Social Considerations Related to VR Technology

Although groundbreaking, VR technology begs various ethical and social issues that demand cautious thought. Given VR systems gather vast user data including physical movements and biometric information, privacy and data security are very critical concerns. If not sufficiently secured, this information could be utilized or abused, therefore compromising major privacy rights. Increasing the risk of psychological effects is VR's immersive character. Users may get dependent on virtual surroundings or lose touch with reality, therefore influencing mental health and social contacts. Dealing with these issues mostly depends on ensuring user privacy and offering strong data protection mechanisms.

The possibility of VR aggravating socioeconomic inequality raises still another important ethical question. Accessibility problems might restrict the advantages of VR for some people, hence extending the digital divide. High prices of VR hardware and software could limit access to people with less means, hence sustaining disparity. Moreover, the material produced for VR might not be inclusive, thus excluding other points of view and needs, so marginalizing some user groups.

Still another area of worry is how VR affects social interactions. Although VR can enable virtual meetings and exchanges, it could potentially lower in-person contact, which is crucial for preserving personal ties and honing interpersonal skills. Dependency too much on VR for social contacts may compromise the quality of real-world contacts, therefore affecting social cohesion and community involvement.

Legal ramifications also follow from VR, especially with regard to intellectual property and content ownership. Defining and safeguarding intellectual property rights for VR experiences get more difficult as they get more sophisticated.

6.5 Challenges and Limitations of VR Technology

Despite its explosive development and expanding uses, VR technology has some important constraints and problems. Hardware restrictions and pricing present one of the main technical difficulties. Excellent VR experiences call for strong computational resources including advanced motion tracking systems, high-resolution monitors, and strong graphics processing. These criteria cause great expenses for consumer acceptance of VR systems as well as for development of them. For many users, the financial obstacle still stands significant, so restricting general acceptance. Furthermore, the demand for significant processing capability could cause hardware failures and overheating, therefore compromising the user experience.

Furthermore, major obstacles in software development are their complexity. Often needing intricate algorithms and real-time rendering, creating immersive and engaging VR experiences calls for both advanced programming and design. Developers have to solve problems including latency, which, if improperly controlled, can induce motion sickness. Including VR technology into current systems and platforms can be difficult, especially when guaranteeing compatibility among several hardware and software environments. Moreover, the need of constant upgrades and problem repairs increases the expenses and burden of development.

One more important area of concern is user experience difficulties. Commonly brought on by mismatched visual and sensory cues, motion sickness, also called VR sickness, is when the VR environment doesn't match users' actual motions, they could get disoriented, nauseous, or dizzy. Another difficult chore is creating user interfaces that are comfortable and easy for long use. Headphones and controllers among other VR gear components depend on their ergonomic design to guarantee user comfort and avoid physical strain.

Further complicating these issues are market and adoption hurdles. Consumer acceptance is still a major barrier notwithstanding developments. The view of VR as a luxury good or a specialist technology can impede its general acceptance. Confusing consumers and restricting interoperability is the scattered character of the VR market with several competing platforms and standards. Table 6.1 outlines key challenges faced in VR technology.

Table 6.1 Challenges in VR technology and potential solutions

Category	Challenge	Impact	Potential solutions
Technical challenges	Hardware limitations and costs	High entry costs, limited accessibility	Reduce production costs, subsidize hardware, and increase component availability
	Software development complexities	Delays, high development costs	Standardize development tools, improve SDKs, and offer developer support
	Integration with existing systems	Implementation delays and increased costs	Develop interoperable systems, provide integration support
User experience challenges	Motion sickness and comfort	Negative user experiences, reduced adoption	Improve latency and refresh rates, enhance ergonomic designs
	User interface design	Poor user engagement and usability	Innovate UI/UX design approaches, conduct user testing
Market and adoption challenges	Consumer acceptance and demand	Slow market growth, low adoption rates	Increase awareness through marketing, reduce costs, offer compelling use cases
	Industry adoption barriers	Slow industry uptake, high investment risk	Provide financial incentives, demonstrate ROI, offer training programs

6.6 Future Trends and Potential Advancements in VR Technology

Driven by continuous improvements across several fields, VR technology's future is set for transforming breakthroughs. Integration of Augmented Reality (AR) and Mixed Reality (MR) with VR is one important development. These technologies hope to produce immersive experiences combining virtual and real worlds, hence improving user interaction and involvement. Whereas MR combines virtual items with physical surroundings in real time, AR overlays digital information over the real world. More dynamic and interactive VR apps where users may easily engage with both virtual and real aspects are predicted from this convergence. Improved by developments in AR and MR, VR worlds will get ever more complex and provide fresh degrees of immersion and utility.

Development of haptic feedback and sensory enhancements presents still another interesting field. While conventional VR systems mostly use visual and aural senses, new technologies are emphasizing enhancing the touch experience. More advanced are haptic feedback systems, which generate physical sensations like force and touch. VR gloves and suits, for example, are in development to replicate textures, impacts, and motions, so users may "feel" virtual objects and environments. These developments should greatly improve the realism and immersion of VR experiences, so transforming virtual interactions into more concrete and understandable form.

With adaptable scenarios and tailored experiences, artificial intelligence may be applied to provide more intelligent and responsive virtual worlds. Real-time material and interactions can be tailored by machine learning algorithms analyzing user preferences and behavior to VR experiences will thus become more tailored and responsive, providing special and interesting material that develops depending on personal user inputs and tastes. AI-driven technologies will also simplify VR content development, hence increasing developer accessibility and allowing the fast creation of intricate virtual environments.

Lighter, more comfortable, and higher-performance headsets are likely to result from VR hardware advances. Higher resolution displays, broader fields of view, and better optics will probably be features of future VR headsets, therefore addressing present limits including screen door effects and visual discomfort. More comfortable users will result from more lightweight and ergonomic designs, therefore enabling longer and more immersive sessions. Better wireless technology will also replace the need for heavy connections, therefore allowing VR experiences to be more flexible and less confined.

Further innovation is expected to be spurred by VR applications' spread over several sectors. New use cases including enhanced simulation training, creative VR experiences, and cooperative VR environments for distant work will surface as VR technology gets more common. The expansion of the VR market will stimulate more research and investment, therefore promoting ongoing developments and fresh VR technological innovations.

6.7 Conclusion

From a niche idea, VR technology has developed quickly into a transforming power in many different sectors. The improvements in VR technology and software have improved user experiences, so realistic and accessible immersive environments are more likely. VR's adaptability and ability to transform established procedures are demonstrated by applications in gaming, education, healthcare, and entertainment. The technology does, however, also bring difficulties including ethical questions, technical restrictions, and obstacles to market acceptance. Driven by developing technologies like augmented reality and advances in artificial intelligence, VR is expected to have notable expansion forward. VR's interaction with other technologies and its spread into other areas will probably inspire creativity and provide fresh ideas to challenging issues as it keeps evolving. Although VR offers fascinating opportunities for the future, it also calls constant attention to solve its problems and make sure it benefits society.

References

1. Abbas, J.R., O'Connor, A., Ganapathy, E., Isba, R., Payton, A., McGrath, B., Tolley, N., Bruce, I.A.: What is virtual reality? A healthcare-focused systematic review of definitions. Health Policy Technol. **12**, 100741 (2023)
2. LaValle, S.M.: Virtual Reality. Cambridge University Press, Cambridge (2023)
3. Kersten, T., Skripnikova, E., Deggim, S.: Development of a virtual reality application for a historic villa and three construction phases. In: Proceedings of the 10th International Workshop 3D-ARCH "3D Virtual Reconstruction and Visualization of Complex Architectures", 21–23 February 2024, Siena, Italy, pp. 247–253 (2024)
4. Aasen, J.: Peer Supported Virtual Reality Development (2023)
5. Mohd, N., Kumar, I.: The importance of virtual reality in the modern and technical Era in computer interaction. In: Proceedings of the 2023 International Conference on Communication, Security and Artificial Intelligence (ICCSAI), pp. 614–618 (2023)
6. Nithva, B., Asha, V., Kumar, K., Giri, J.: Efficacious opportunities and implications of virtual reality features and techniques. In: Proceedings of the 2022 Fourth International Conference on Cognitive Computing and Information Processing (CCIP), pp. 1–8 (2022)
7. Dobhal, D.C., Negi, H.S., Das, P.: The effect of virtual reality in the modern system of education. In: Proceedings of the 2023 2nd International Conference for Innovation in Technology (INOCON), pp. 1–4 (2023)
8. Zhong, Z.: The impact of virtual reality on gaming. Highl. Sci. Eng. Technol. **93**, 179–184 (2024)
9. Mao, Z., Wong, C., Zhang, Y.: Overview of the application of virtual reality in the field of visual arts and video games. Highl. Sci. Eng. Technol. **44**, 333–340 (2023)
10. Jeon, S., Yang, X.: Application of VR In game design & filmmaking. Highl. Sci. Eng. Technol. **41**, 371–376 (2023)
11. Kamińska, D., Sapiński, T., Wiak, S., Tikk, T., Haamer, R.E., Avots, E., Helmi, A., Ozcinar, C., Anbarjafari, G.: Virtual reality and its applications in education: survey. Information. **10**, 318 (2019)
12. Jiang, J., Zhi, L., Xiong, Z.: Application of virtual reality technology in education and teaching. In: Proceedings of the 2018 International joint conference on information, media and engineering (ICIME), pp. 300–302 (2018)
13. Lin, Y., You, Y., Zhou, Z.: Application of VR in the field of education. Highl. Sci. Eng. Technol. **68**, 11–20 (2023)
14. Javaid, M., Haleem, A.: Virtual reality applications toward medical field. Clin. Epidemiol. Global Health. **8**, 600–605 (2020)
15. Samadbeik, M., Yaaghobi, D., Bastani, P., Abhari, S., Rezaee, R., Garavand, A.: The applications of virtual reality technology in medical groups teaching. J. Adv. Med. Educ. Prof. **6**, 123 (2018)

Chapter 7
Quantum Computing

7.1 Introduction to Quantum Computing

Quantum computing represents a revolutionary shift from traditional, or classical, computing by leveraging principles of quantum mechanics to process information in fundamentally new ways. While classical computers rely on bits that exist as either 0 s or 1 s, quantum computers use quantum bits, or qubits, which can exist simultaneously in multiple states due to a phenomenon known as superposition. This capability allows quantum computers to process complex calculations at speeds that far exceed those of classical computers, particularly for certain types of problems. Entanglement, another principle of quantum mechanics, further amplifies this power by allowing qubits to be interconnected in ways that make computations more efficient. The difference between these two computing paradigms sets the stage for understanding the transformative potential of quantum technology [1].

The implications of quantum computing extend across numerous fields and industries. In finance, for example, quantum computers could optimize vast portfolios and manage financial risk at unprecedented scales. In healthcare, quantum computing holds the promise of accelerating drug discovery by simulating molecular interactions with high precision. Supply chains, logistics, and even renewable energy sectors also stand to benefit from the unique problem-solving capabilities that quantum computing brings. The possibilities are profound, yet they also raise critical questions about the potential impacts on current technology infrastructures and society at large [1].

This chapter aims to introduce the fundamental principles of quantum computing, discuss the core technologies driving its development, and explore some of the most promising algorithms that distinguish quantum from classical approaches. By examining practical applications, current challenges, and ethical considerations, this chapter will provide a comprehensive overview of quantum computing and its

H. Taherdoost, *Innovating and Integrating Emerging Technologies*, Signals and Communication Technology, https://doi.org/10.1007/978-3-031-93953-2_7

potential trajectory. In doing so, it will address the transformative role that quantum computing could play in shaping the future of technology and the digital landscape.

7.2 Fundamentals of Quantum Computing

Quantum computing is built on the foundation of quantum bits, or qubits, which differ significantly from the binary bits used in classical computing (Fig. 7.1). While classical bits can represent only one of two possible states, either 0 or 1, qubits can exist in a superposition of states, allowing them to represent both 0 and 1 simultaneously. This unique property of superposition enables quantum computers to process vast amounts of information at once, dramatically increasing computational power for certain tasks compared to classical computers.

Entanglement, another core concept of quantum mechanics, enhances this computational advantage. When qubits become entangled, the state of one qubit becomes intrinsically linked to the state of another, regardless of the distance between them. This means that any change in the state of one entangled qubit will instantaneously affect the other, a phenomenon that classical computers cannot replicate. Quantum interference, a third principle, enables quantum systems to amplify the probability of correct answers while canceling out incorrect ones during calculations. These principles, superposition, entanglement, and interference, are central to the power and efficiency of quantum computing, enabling it to tackle problems that are practically unsolvable for classical computers.

At the core of quantum computing lies the concept of quantum bits, or qubits, which represent a fundamental departure from the binary bits used in classical computing. Classical bits can exist in only one of two states, either 0 or 1, limiting them to a linear form of computation. Qubits, on the other hand, can exist in a state of superposition, meaning they can represent both 0 and 1 simultaneously. This characteristic allows quantum computers to handle multiple possibilities at once,

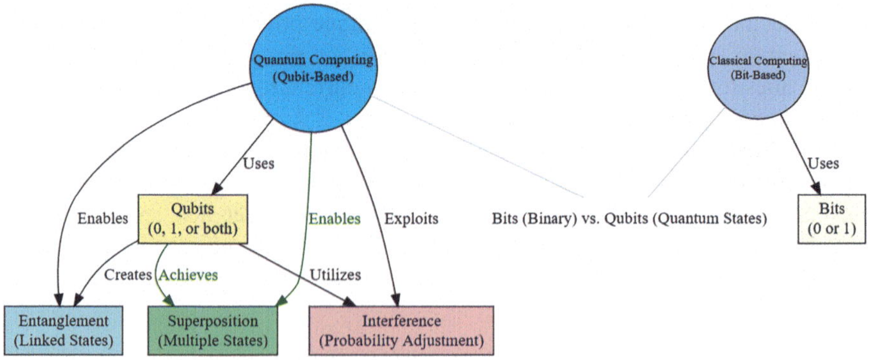

Fig. 7.1 Core concepts in quantum computing

exponentially increasing their computational power for certain types of problems. Superposition enables a quantum computer to perform parallel calculations, making it capable of processing complex data in ways that are out of reach for classical systems [1, 2].

Entanglement is another powerful property of quantum computing, which distinguishes it from traditional computing models. When two qubits become entangled, the state of one qubit becomes intrinsically linked to the state of the other, regardless of the physical distance between them. This interdependence means that any change to one qubit's state will immediately affect the entangled partner, a feature that classical computers cannot replicate. Entanglement enhances computational capabilities by allowing qubits to work in tandem, creating a networked system where changes propagate rapidly and in synchrony.

Another essential concept in quantum computing is interference, which harnesses the probabilistic nature of quantum mechanics to refine computational outcomes. Interference helps quantum systems increase the likelihood of correct answers by enhancing the probabilities associated with accurate solutions while canceling out the probabilities of incorrect ones. Quantum gates and circuits are the operational building blocks that enable quantum computers to execute tasks using these unique properties. Quantum gates manipulate qubits' states in ways that exploit superposition, entanglement, and interference, forming circuits that are fundamentally different from classical logic circuits. The combined use of superposition, entanglement, and interference, along with quantum gates, allows quantum computers to perform calculations with a speed and complexity that redefines traditional computing limits.

7.3 Quantum Computing Hardware

Quantum computing hardware is driven by the foundational component known as the qubit, a quantum equivalent of the classical bit, but capable of representing both 0 and 1 simultaneously through superposition (Fig. 7.2). Different types of qubits have been developed, each with unique advantages and challenges that make them suitable for various quantum computing architectures. Among the most prominent qubit types are superconducting qubits, ion trap qubits, and photonic qubits. Superconducting qubits, which operate at temperatures near absolute zero, leverage superconducting circuits to create stable quantum states. They are widely used by companies like IBM and Google due to their relatively straightforward integration into existing hardware infrastructure, although they require extremely cold environments to function effectively. Ion trap qubits, on the other hand, use ions suspended in electromagnetic fields. These qubits offer long coherence times and high fidelity but require sophisticated control and isolation from external disturbances. Photonic qubits, based on particles of light, have the advantage of operating at room temperature, making them suitable for quantum communication networks. However,

Fig. 7.2 Current hardware development in quantum computing

manipulating and maintaining photonic qubits with precision present significant engineering challenges.

The development of quantum computing hardware is advancing rapidly, with several leading organizations at the forefront of this technological race. IBM, Google, and Rigetti Computing are among the key players developing quantum processors and pioneering advances in qubit technology. IBM has established its quantum computing platform, IBM Quantum, and offers access to its quantum processors through cloud services. IBM's approach to scaling quantum hardware emphasizes both research in superconducting qubits and building a quantum ecosystem accessible to researchers and developers globally. Google's quantum computing research, known for achieving a milestone with "quantum supremacy" in 2019, focuses on using superconducting qubits to solve computational problems beyond the reach of classical computers. Rigetti Computing, a smaller but influential player, has concentrated on developing full-stack quantum computing solutions, combining hardware with software tailored for quantum applications. These companies are in a continuous race to develop more stable, powerful, and scalable quantum computers that can support practical applications [3].

Quantum hardware development, however, faces significant challenges, primarily related to error rates, coherence times, and scalability. Quantum systems are exceptionally sensitive to environmental noise, which results in errors during computation. Error rates remain high in quantum processors, making error correction a critical research area in quantum computing. Quantum error correction techniques are being explored to mitigate these errors, but they require additional qubits and computational resources, adding complexity to hardware design. Coherence time, the duration for which a qubit can maintain its quantum state, is another challenge. Quantum states are inherently fragile, and maintaining coherence over longer periods is essential for performing complex calculations. Advances in materials science, cryogenics, and quantum error correction are critical for addressing this limitation. Finally, scalability is a formidable challenge; building quantum computers with thousands or millions of qubits is necessary for practical applications, yet creating and managing large arrays of qubits remain a considerable hurdle. Addressing these hardware challenges is crucial as quantum computing moves from experimental stages toward practical, large-scale applications, and it will determine the pace and success of quantum computing as a transformative technology.

7.4 Quantum Algorithms

Quantum algorithms are at the heart of quantum computing, leveraging the unique properties of quantum mechanics to solve problems that are either impossible or computationally expensive for classical computers. Notable quantum algorithms include Shor's algorithm and Grover's algorithm, which have demonstrated the potential of quantum computers to surpass classical systems in specific applications. Shor's algorithm, developed by mathematician Peter Shor in 1994, is particularly important in the field of cryptography [4]. It offers an exponential speedup in factoring large integers, a problem that forms the basis of many encryption schemes, such as RSA. This breakthrough suggests that, with sufficiently advanced quantum computers, classical encryption methods could be compromised, sparking an ongoing exploration of quantum-resistant encryption algorithms.

Grover's algorithm, introduced by Lov Grover in 1996, provides a quantum speedup for unstructured search problems. It is particularly effective for searching through large databases or solving optimization problems, where classical methods would require linear time to check each possibility. Grover's algorithm achieves a quadratic speedup, making it an attractive approach for tasks involving searching, such as data mining, and optimization problems that arise in various fields like logistics, machine learning, and artificial intelligence [5].

Recent advances in quantum computing have also spurred interest in applications beyond basic algorithms. Quantum machine learning, for instance, combines quantum computing with machine learning techniques to potentially offer improvements in processing large datasets and training complex models. Quantum-enhanced machine learning algorithms could revolutionize fields such as natural language processing, image recognition, and predictive analytics. Similarly, quantum cryptography holds the promise of creating ultra-secure communication systems, utilizing quantum key distribution (QKD) protocols to ensure the privacy of transmitted information in a way that is theoretically impossible for classical computers to breach. Furthermore, quantum computing is being explored for its ability to solve complex optimization problems, such as those encountered in supply chain management, financial portfolio optimization, and drug discovery. Table 7.1 summarizing notable quantum algorithms, their primary applications, and their potential impact.

Table 7.1 Notable quantum algorithms and their applications

Algorithm	Primary application	Potential impact
Shor's algorithm	Factorization of large integers	Breaks RSA encryption, leading to the need for quantum-resistant cryptographic techniques
Grover's algorithm	Unstructured search, optimization problems	Provides quadratic speedup for searching databases and solving optimization problems
Quantum machine learning	Data processing, training models	Enhances machine learning techniques for faster data processing and improved model training
Quantum cryptography	Secure communication (quantum key distribution)	Ensures theoretically unbreakable encryption for communication, enhancing privacy and security

7.5 Quantum Computing Applications Across Industries

Quantum computing is poised to revolutionize multiple industries by providing solutions to complex problems that are currently beyond the capabilities of classical computers (Fig. 7.3). In finance, for instance, quantum computing has the potential to significantly improve portfolio optimization and risk analysis. Quantum algorithms can process vast amounts of data much faster than traditional systems, allowing for more accurate financial modeling and forecasting. This can help financial institutions make more informed investment decisions and optimize trading strategies. Moreover, the ability to efficiently solve complex problems such as risk management and pricing derivatives is expected to transform the landscape of financial services, giving firms a competitive edge in an increasingly data-driven market.

In healthcare, quantum computing could accelerate drug discovery and revolutionize genomics research. By simulating molecular interactions at an unprecedented level of detail, quantum computers can help scientists identify promising drug candidates much faster and with greater precision than classical methods allow. This could drastically reduce the time and cost associated with bringing new drugs to market. Additionally, quantum computing can aid in understanding complex genetic data, leading to breakthroughs in personalized medicine and the treatment of genetic disorders. Quantum algorithms can potentially enhance the ability to analyze and manipulate genomic data, facilitating advances in gene editing and disease prevention.

The logistics industry also stands to benefit greatly from quantum computing, particularly in the areas of route optimization and supply chain management. Quantum algorithms can analyze multiple variables and constraints simultaneously, making it possible to find the most efficient routes for delivery fleets, even in highly dynamic environments. This can lead to significant reductions in transportation costs, energy usage, and delivery times. Additionally, quantum computing can improve supply chain management by optimizing inventory control, predicting demand fluctuations, and enhancing the overall efficiency of production processes.

Fig. 7.3 Quantum computing applications across industries

These improvements could result in more sustainable and cost-effective logistics operations.

In the energy sector, quantum computing has the potential to improve modeling and simulation processes, leading to more efficient energy production and distribution. By simulating complex systems like electrical grids and renewable energy networks, quantum computers can help design more efficient systems that maximize energy output while minimizing waste. Quantum algorithms could also play a key role in optimizing energy storage solutions, which is critical for the adoption of renewable energy sources. In addition, the ability to model molecular interactions in energy production, such as in the development of new materials for batteries or fuel cells, could lead to significant advancements in energy efficiency. As quantum computing technology matures, its applications in energy, healthcare, finance, and logistics will likely drive innovation and contribute to the development of more sustainable and efficient systems across these industries.

7.6 Challenges and Limitations

Quantum computing, while promising immense potential, faces several challenges and limitations that must be addressed before it can become widely applicable. One of the primary technical challenges is quantum decoherence, which occurs when quantum bits (qubits) lose their quantum properties due to interaction with their environment. This loss of coherence significantly hampers the ability of quantum computers to perform accurate and reliable computations. Alongside this, error correction remains a major hurdle. Unlike classical computers, where errors can be detected and corrected using straightforward methods, quantum computers require complex error correction techniques that can handle the fragile nature of qubits. Developing robust and scalable error correction methods is essential for ensuring that quantum computers can function at scale and solve real-world problems.

In addition to the technical hurdles, quantum computing also has significant resource and infrastructure requirements. Quantum processors require highly controlled environments, often operating at extremely low temperatures to maintain the qubits' quantum state. The infrastructure needed to support these conditions, such as dilution refrigerators and specialized shielding, is both costly and complex. Moreover, quantum computers demand an immense amount of computational power, with even modest quantum systems requiring a substantial number of qubits and resources to carry out meaningful tasks. As quantum technology progresses, the need for further advancements in hardware and supportive infrastructure will only increase, making it a significant barrier to widespread deployment.

Beyond technical and infrastructure challenges, ethical and security concerns present another critical area of focus. As quantum computing has the potential to break current encryption methods, it raises significant concerns about data protection. Classical encryption algorithms, such as RSA, rely on the difficulty of factoring large numbers, a problem that quantum computers can solve efficiently using algorithms like Shor's algorithm. This creates the risk of compromising sensitive data stored or transmitted using current cryptographic standards. As a result, there is an urgent need for the development of quantum-resistant encryption algorithms that can protect data against potential quantum-based threats. Moreover, the ethical implications of quantum computing extend beyond data security, encompassing issues such as access to quantum technologies, their impact on privacy, and the potential for misuse in harmful ways. These concerns must be carefully considered as the technology evolves to ensure that quantum computing benefits society as a whole while minimizing risks.

7.7 Future of Quantum Computing

The future of quantum computing holds significant promise, though its full potential will unfold over both the short and long terms. In the short term, experts predict breakthroughs in quantum error correction, which will enhance the reliability of quantum systems. As researchers continue to overcome challenges related to qubit coherence and error rates, more stable and scalable quantum computers may emerge. Another area of anticipated progress is the development of hybrid computing models, where quantum processors are integrated with classical systems to leverage the strengths of both. This integration could accelerate the practical application of quantum computing in industries such as finance, healthcare, and logistics by providing powerful tools that can tackle specific tasks beyond the reach of classical computers. In the coming decade, improvements in quantum hardware and algorithms are expected to make quantum computers more accessible and capable of solving complex problems that current technologies cannot handle.

Long-term potential for quantum computing includes its transformative impact on fields like cryptography, artificial intelligence, and materials science. Quantum algorithms have the capacity to revolutionize optimization problems, simulate molecular structures for drug discovery, and provide breakthroughs in machine learning. With time, as quantum computers achieve greater stability and scalability, they could vastly enhance computational efficiency, solving problems in minutes that would otherwise take classical supercomputers years. This transformation could reshape industries by providing solutions to problems once thought insurmountable, opening doors to innovations that will influence the future of technology and society.

Quantum computing will likely integrate more seamlessly with other technologies, leading to new hybrid computing models. Quantum cloud services, where users can access quantum computing power remotely, are already on the horizon, making this powerful technology more accessible to a wider audience. By combining the power of quantum computing with the versatility of cloud infrastructure, businesses and research institutions could harness the potential of quantum systems without the need for massive investments in quantum hardware. This shift toward hybrid computing could facilitate rapid advancements in AI, optimization, and data analytics, benefiting various sectors from finance to healthcare and beyond.

7.8 Conclusion

In conclusion, the development of quantum computing is still in its early stages, but its potential is vast and transformative. As breakthroughs in quantum hardware and algorithms unfold, the possibilities for how quantum computing can impact our world continue to expand. In the short term, integration with classical systems and improvements in error correction will enhance its usability, while in the long term, quantum computing is poised to revolutionize a wide range of industries. As this journey progresses, the future of technology and society will likely be shaped by the capabilities of quantum systems, ushering in an era of computing that we are only beginning to understand.

References

1. Hirvensalo, M.: Quantum computing. Springer Science & Business Media, Cham (2013)
2. Clarke, J., Wilhelm, F.K.: Superconducting quantum bits. Nature. **453**(7198), 1031–1042 (2008)
3. Singh, P., Dasgupta, R., Singh, A., Pandey, H., Hassija, V., Chamola, V., Sikdar, B.: A survey on available tools and technologies enabling quantum computing. IEEE Access (2024)
4. Ugwuishiwu, C.H., Orji, U.E., Ugwu, C.I., Asogwa, C.N.: An overview of quantum cryptography and Shor's algorithm. Int. J. Adv. Trends Comput. Sci. Eng. **9**(5), 7487–7495 (2020)
5. Khurana, S., Nene, M. J.: Implementation of database search with quantum computing: Grover's Algorithm vs Linear Search. In: 2023 International Conference on Ambient Intelligence, Knowledge Informatics and Industrial Electronics (AIKIIE), IEEE, pp. 1–6 (2023, November)

Chapter 8
Cybersecurity Frontiers

8.1 Introduction

Cybersecurity is a critical field that encompasses the protection of computer systems, networks, and data from malicious attacks, unauthorized access, and damage. As our reliance on digital technologies increases, so does the importance of cybersecurity in safeguarding sensitive information and maintaining the integrity of our digital infrastructure.

Cybersecurity can be defined as the practice of implementing measures to protect internet-connected devices and services from cyber threats such as hacking, phishing, and ransomware attacks. It involves a combination of technologies, processes, and practices designed to defend against these threats and ensure the confidentiality, integrity, and availability of data. The significance of cybersecurity extends beyond mere data protection; it is essential for maintaining trust in digital interactions, ensuring compliance with regulations, and protecting the financial and reputational interests of organizations and individuals alike. The consequences of inadequate cybersecurity can be severe, including financial losses, legal repercussions, and damage to brand reputation [1].

The digital landscape today is characterized by an interconnected web of devices and systems, ranging from personal computers and smartphones to cloud services and the Internet of Things (IoT). This interconnectedness, while facilitating convenience and efficiency, also creates a vast attack surface for cybercriminals. Each device and application represents a potential vulnerability that can be exploited. The rise of remote work, accelerated by the COVID-19 pandemic, has further complicated the security landscape, as employees access corporate networks from various locations and devices, often without adequate security measures in place. In this environment, the potential for cyberattacks is ever-increasing. Cybercriminals employ sophisticated techniques to exploit vulnerabilities, leading to data breaches, identity theft, and financial fraud. The challenge for organizations is to implement

H. Taherdoost, *Innovating and Integrating Emerging Technologies*, Signals and Communication Technology, https://doi.org/10.1007/978-3-031-93953-2_8

comprehensive cybersecurity strategies that address these vulnerabilities while enabling the benefits of digital transformation. This includes adopting robust security protocols, conducting regular risk assessments, and fostering a culture of cybersecurity awareness among employees.

The history of cybersecurity is marked by the evolution of cyber threats and the corresponding development of defensive measures (Fig. 8.1). The concept of cybersecurity began to take shape in the early 1970s with the advent of ARPANET, the precursor to the internet. Initial efforts focused on securing mainframe computers and protecting sensitive military information. However, as the internet became widely accessible in the 1990s, the landscape of cyber threats expanded dramatically. Notable incidents, such as the Morris Worm in 1988, the first computer worm distributed via the internet, highlighted the vulnerabilities inherent in interconnected systems. The late 1990s and early 2000s saw the emergence of more sophisticated attacks, including the Melissa virus and the ILOVEYOU worm, which caused significant disruptions and financial losses. The 2010s introduced advanced persistent threats (APTs) and state-sponsored cyberattacks, further complicating the cybersecurity landscape. Today, organizations face a myriad of threats, including ransomware attacks that can cripple operations and data breaches that expose sensitive customer information. The increasing frequency and severity of these incidents underscore the need for robust cybersecurity measures and a proactive approach to threat management. As technology continues to evolve, so too will the tactics employed by cybercriminals, making cybersecurity an ongoing and dynamic challenge for individuals and organizations alike [2].

8.2 Emerging Cyber Threats

Emerging cyber threats are increasingly complex and varied, reflecting the rapid evolution of technology and the tactics employed by cybercriminals (Fig. 8.2). Understanding these threats is essential for organizations and individuals seeking to protect their digital assets. Malware encompasses various malicious software, including viruses, worms, and Trojans, designed to disrupt, damage, or gain unauthorized access to computer systems. It can be delivered through infected email attachments, compromised websites, or removable media. Phishing attacks involve deceptive emails or messages that appear legitimate, tricking users into revealing sensitive information such as passwords or credit card numbers. Recent trends show that attackers are using artificial intelligence to create more convincing phishing schemes, increasing their effectiveness. Ransomware is a particularly insidious type of malware that encrypts a victim's files, demanding a ransom for the decryption key. This threat has surged in recent years, with high-profile attacks affecting critical infrastructure and major corporations, often resulting in significant financial losses. Advanced Persistent Threats (APTs) are prolonged and targeted cyberattacks where an intruder gains access to a network and remains undetected for an extended period [3]. These attacks are often state-sponsored and aim to steal sensitive data or

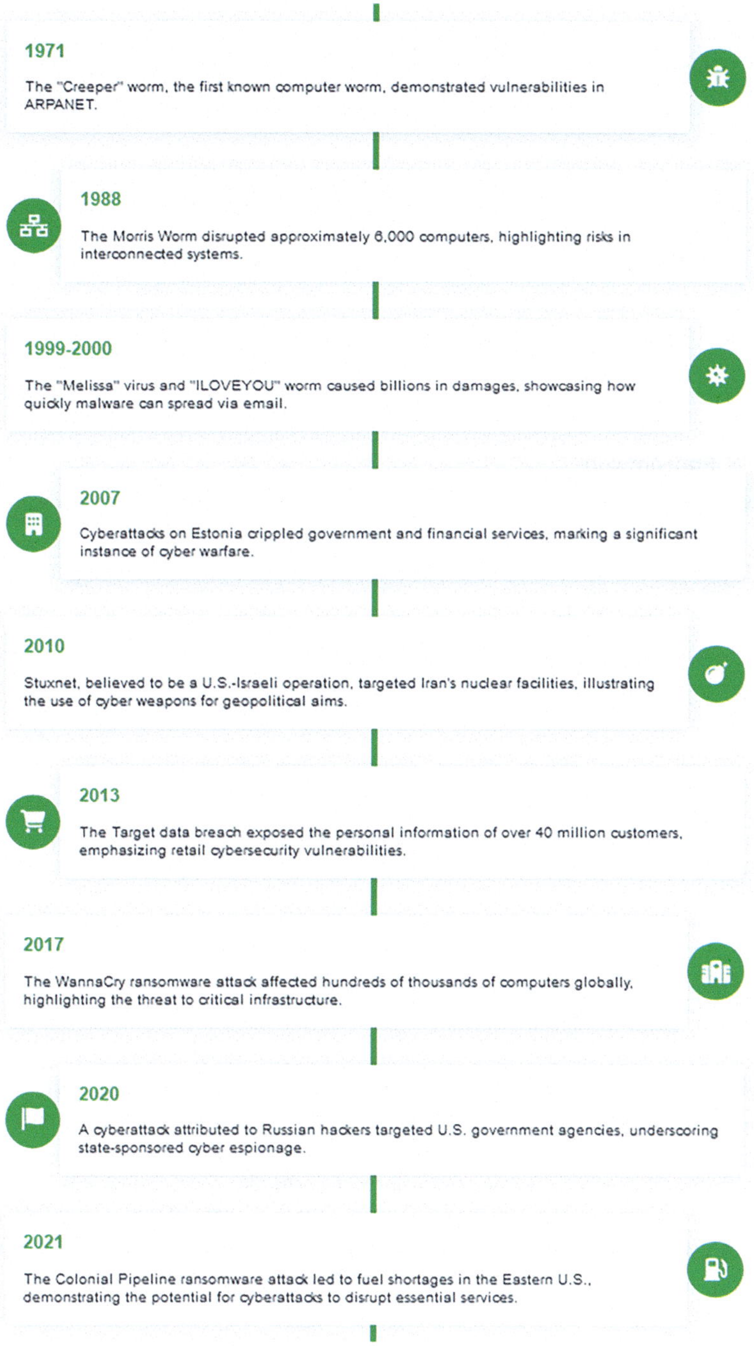

Fig. 8.1 Historical context of cyber threats and attacks

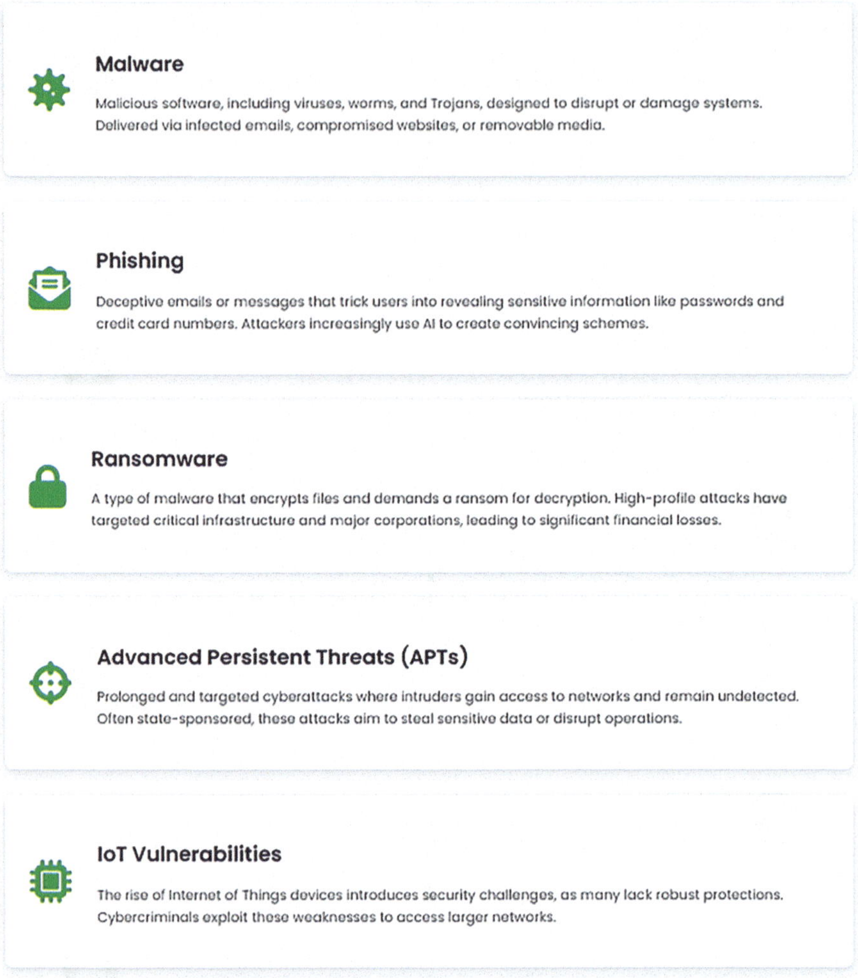

Fig. 8.2 Emerging cyber threats overview

disrupt operations. Additionally, the proliferation of Internet of Things (IoT) devices introduces new security challenges. Many IoT devices lack robust security features, making them attractive targets for cybercriminals who can exploit these weaknesses to launch attacks or gain access to larger networks. Several significant cyber incidents illustrate the growing complexity of these threats. The WannaCry ransomware attack in 2017 affected hundreds of thousands of computers across 150 countries, exploiting a vulnerability in Microsoft Windows. It disrupted services in various sectors, including healthcare, where it caused significant delays in patient care. The SolarWinds supply chain attack in 2020 involved the compromise of the SolarWinds Orion software, affecting numerous U.S. government agencies and private companies [4]. Attackers inserted malicious code into legitimate software updates, highlighting vulnerabilities in supply chain security. The Colonial Pipeline ransomware attack in 2021 led to fuel supply disruptions across the Eastern United States [5]. The company paid a ransom of approximately $4.4 million to regain access to its systems, emphasizing the real-world impact of cyber threats on critical infrastructure. Emerging technologies, particularly artificial intelligence (AI) and the Internet of Things (IoT), are reshaping the cybersecurity landscape. While AI enhances cybersecurity defenses by improving threat detection and response times, it also empowers cybercriminals. Attackers can use AI to automate attacks, analyze vulnerabilities, and craft more convincing phishing messages. The dual-use nature of AI necessitates a proactive approach to cybersecurity that anticipates and mitigates these risks. The rapid expansion of IoT devices has created numerous entry points for cyberattacks. Many IoT devices are inadequately secured, making them susceptible to exploitation. As these devices become more integrated into everyday life and critical infrastructure, the potential for widespread disruption grows.

8.3 Cybersecurity Principles

Cybersecurity is built upon a foundation of key principles that guide organizations in protecting their digital assets. At the core of these principles lies the CIA triad, which encompasses Confidentiality, Integrity, and Availability. Additionally, effective cybersecurity requires a proactive approach to risk management and assessment, as well as the implementation of robust security policies and frameworks. The CIA triad is a widely recognized model that serves as the cornerstone of information security. Each element of the triad plays a crucial role in safeguarding data and systems. Confidentiality ensures that sensitive information is accessible only to authorized individuals or entities. This principle involves implementing measures such as access controls, encryption, and secure communication protocols to prevent unauthorized disclosure of data. Integrity maintains the accuracy, completeness, and trustworthiness of data throughout its entire lifecycle. It ensures that information is not altered or tampered with by unauthorized parties and that it remains reliable for decision-making and operations. Availability guarantees that authorized users have reliable and timely access to information and resources when needed. This principle involves implementing measures to prevent disruptions, minimize downtime, and ensure the continuous operation of critical systems and services. By adhering to the CIA triad, organizations can develop a comprehensive approach to cybersecurity that addresses the fundamental aspects of data protection.

Effective cybersecurity requires a proactive approach to risk management and assessment. Organizations must identify, analyze, and mitigate potential threats and vulnerabilities that could compromise their digital assets (Fig. 8.3). This process begins with risk identification, where potential threats, both internal and external, are recognized. This includes considering factors such as threat actors, their motivations, and the potential attack vectors. Following identification, risk analysis assesses the likelihood and potential impact of these risks, helping to prioritize them

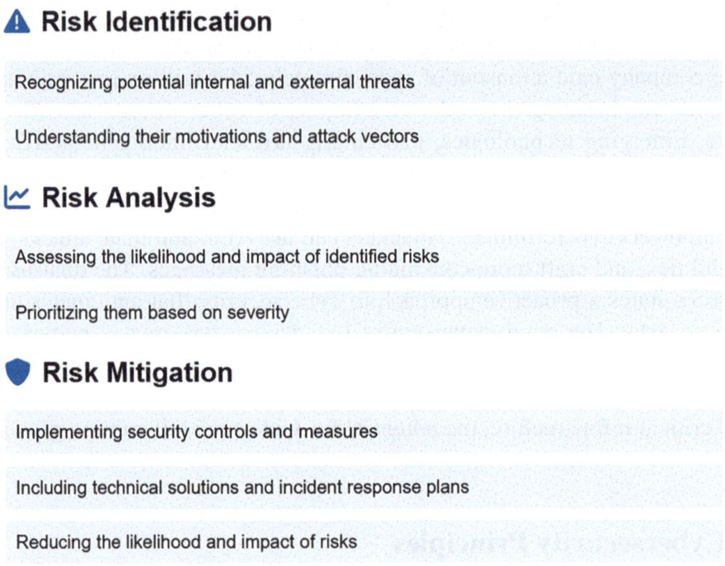

Fig. 8.3 Risk management and assessment

based on severity and guiding the allocation of resources for mitigation efforts. Risk mitigation involves implementing appropriate security controls and measures to reduce the likelihood and impact of identified risks. This may include technical solutions, such as firewalls and intrusion detection systems, as well as developing incident response plans and conducting regular security assessments. By continuously assessing and managing risks, organizations can proactively address security vulnerabilities and minimize the potential for successful attacks.

To ensure the consistent and effective implementation of cybersecurity measures, organizations should establish comprehensive security policies and frameworks (Fig. 8.4). These policies define the organization's security objectives, roles and responsibilities, and expected behaviors from employees and third parties. Security frameworks, such as the NIST Cybersecurity Framework or ISO/IEC 27001, provide a structured approach to managing information security risks and implementing best practices [6]. Key elements of security policies and frameworks include asset management, which involves identifying and classifying the organization's digital assets to prioritize security efforts. Access control measures ensure that only authorized individuals or entities can access and interact with the organization's systems and data. Incident response procedures establish protocols for detecting, responding to, and recovering from security incidents, minimizing the impact on the organization's operations and reputation. Compliance with relevant laws, regulations, and industry standards, such as GDPR, HIPAA, or PCI-DSS, is essential to mitigate legal and financial risks.

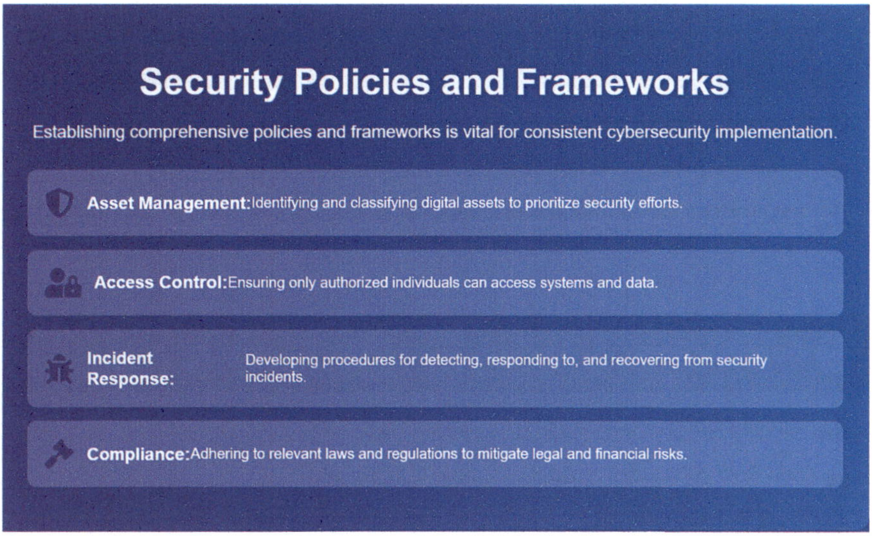

Fig. 8.4 Security policies and frameworks

8.4 Cyber Defense Mechanisms

Cyber defense mechanisms are essential for protecting organizations against the increasing frequency and sophistication of cyber threats (Fig. 8.5). These mechanisms encompass a range of defensive technologies, best practices for securing networks and data, and strategies for incident response and disaster recovery. Defensive technologies form the first line of defense against cyber threats. Firewalls are critical components that monitor and control incoming and outgoing network traffic based on predetermined security rules. They act as barriers between trusted internal networks and untrusted external networks, preventing unauthorized access and potential attacks. Firewalls can be hardware-based, software-based, or a combination of both, and they are essential for establishing a secure perimeter around an organization's network. Antivirus software is another vital defensive technology that detects, prevents, and removes malicious software, including viruses, worms, and Trojans. Modern antivirus solutions employ heuristic analysis and machine learning to identify new and evolving threats, providing real-time protection against malware. Regular updates to virus definitions are crucial to ensure that the antivirus software can recognize the latest threats. Intrusion Detection System (IDS) and Intrusion Prevention System (IPS) are also integral to cybersecurity. An IDS monitors network traffic for suspicious activity and alerts administrators to potential threats, while an IPS takes this a step further by actively blocking or preventing those threats. These systems analyze traffic patterns and can identify anomalies that may indicate an attack, allowing organizations to respond swiftly to potential breaches. In addition to these technologies, best practices for securing networks and

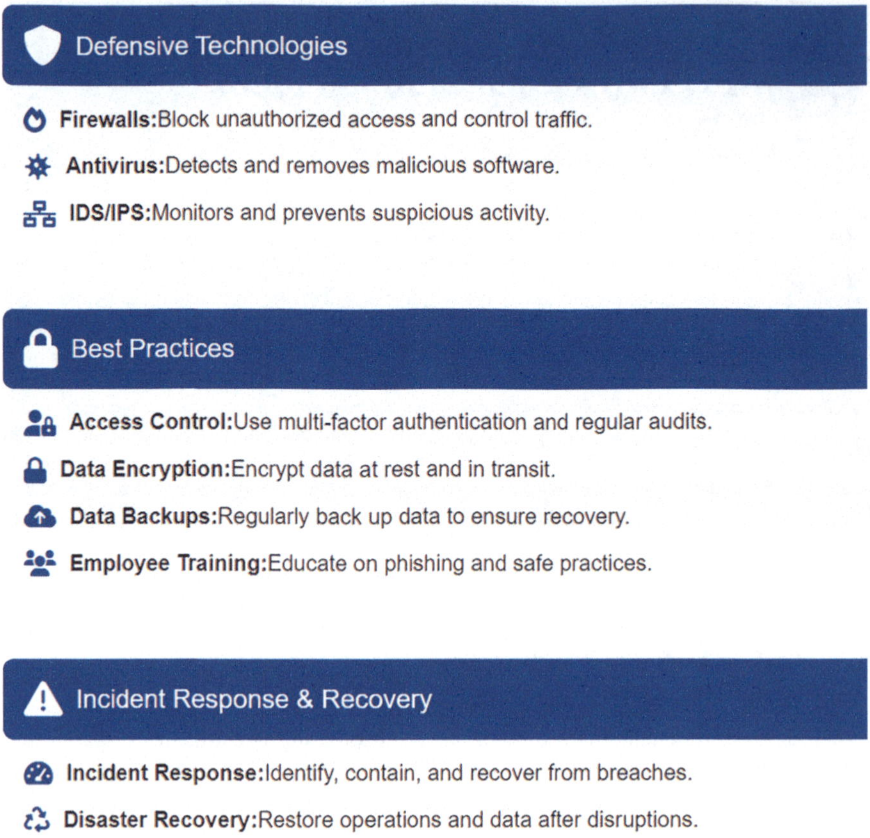

Fig. 8.5 Cyber defense mechanisms overview

data are fundamental to a robust cybersecurity posture. Organizations should implement strong access control measures, ensuring that only authorized personnel have access to sensitive information and systems. This can be achieved through multi-factor authentication, role-based access controls, and regular audits of user permissions. Data encryption is another critical practice that protects sensitive information by converting it into an unreadable format for unauthorized users. Encryption should be applied to data at rest and in transit, ensuring that even if data is intercepted, it remains secure. Regular data backups are also essential, allowing organizations to recover information in the event of a ransomware attack or data loss incident. Employee training and awareness programs are vital components of a comprehensive cybersecurity strategy. Human error is often a significant factor in successful cyberattacks, so educating employees about phishing scams, social engineering tactics, and safe internet practices can significantly reduce the risk of breaches. Regular security drills and updates on emerging threats can help keep cybersecurity at the forefront of organizational culture. Incident response strategies and disaster recovery plans are crucial for minimizing the impact of cyber incidents when they occur. An effective incident response plan outlines the steps to be taken in the event of a security breach, including identifying the source of the breach, containing the threat, eradicating it, and recovering affected systems. This plan should include clear roles and responsibilities for team members, ensuring a coordinated response. Disaster recovery plans complement incident response strategies by focusing on restoring operations after a significant disruption. These plans should include detailed procedures for data recovery, system restoration, and business continuity. Regular testing of these plans is essential to ensure their effectiveness and to identify areas for improvement.

8.5 Legal and Ethical Considerations

Cyber defense mechanisms are crucial for protecting organizations against the ever-increasing frequency and sophistication of cyber threats (Table 8.1). These mechanisms encompass a variety of defensive technologies, best practices for securing networks and data, and strategies for incident response and disaster recovery. Defensive technologies form the first line of protection against cyber threats. Firewalls are essential components that monitor and control incoming and outgoing network traffic based on predetermined security rules. Acting as barriers between trusted internal networks and untrusted external networks, firewalls prevent unauthorized access and potential attacks. They can be hardware-based, software-based, or a combination of both, and they are vital for establishing a secure perimeter around an organization's network. Antivirus software is another critical defensive technology that detects, prevents, and removes malicious software, including viruses, worms, and Trojans. Modern antivirus solutions employ heuristic analysis and machine learning to identify new and evolving threats, providing real-time protection against malware. Regular updates to virus definitions are necessary to ensure that the antivirus software can recognize the latest threats and effectively combat them. Intrusion Detection System (IDS) and Intrusion Prevention System (IPS) are also integral to a comprehensive cybersecurity strategy. An IDS monitors network traffic for suspicious activity and alerts administrators to potential threats, while an IPS goes a step further by actively blocking or preventing those threats. These systems analyze traffic patterns and can identify anomalies that may indicate an attack, allowing organizations to respond swiftly to potential breaches. In addition to these technologies, implementing best practices for securing networks and data is fundamental to a robust cybersecurity posture [7]. Organizations should enforce strong access control measures to ensure that only authorized personnel have access to sensitive information and systems. This can be achieved through multi-factor authentication, role-based access controls, and regular audits of user permissions. Data encryption is another critical practice that protects sensitive information by converting it into an unreadable format for unauthorized users. Encryption should be applied to data both at rest and in transit, ensuring that even if data is intercepted, it remains secure. Regular data backups are also essential for maintaining data

Table 8.1 Comprehensive cyber defense mechanisms

Category	Mechanism/Practice	Description
Defensive technologies	Firewalls	Monitor and control incoming and outgoing network traffic based on predetermined security rules
	Antivirus software	Detects, prevents, and removes malicious software, employing heuristic analysis and machine learning
	Intrusion detection system (IDS)	Monitors network traffic for suspicious activity and alerts administrators to potential threats
	Intrusion prevention system (IPS)	Actively blocks or prevents identified threats, analyzing traffic patterns for anomalies
Best practices for security	Access control measures	Ensures that only authorized personnel can access sensitive information through multi-factor authentication and role-based access
	Data encryption	Protects sensitive information by converting it into an unreadable format for unauthorized users
	Regular data backups	Facilitates recovery of information in case of ransomware attacks or data loss incidents
	Employee training and awareness	Educates employees about phishing scams, social engineering, and safe internet practices to reduce human error
Incident response strategies	Incident response plan	Outlines steps to take during a security breach, including identification, containment, eradication, and recovery
	Roles and responsibilities	Clearly defines team member roles to ensure a coordinated response during an incident
Disaster recovery	Disaster recovery plan	Focuses on restoring operations after a significant disruption, including data recovery and system restoration
	Regular testing of plans	Ensures the effectiveness of incident response and disaster recovery plans, identifying areas for improvement

integrity and availability. By backing up data frequently, organizations can recover information in the event of a ransomware attack or data loss incident, minimizing the impact on operations. Furthermore, employee training and awareness programs are vital components of a comprehensive cybersecurity strategy. Human error is often a significant factor in successful cyberattacks, so educating employees about phishing scams, social engineering tactics, and safe internet practices can significantly reduce the risk of breaches. Regular security drills and updates on emerging threats can help keep cybersecurity at the forefront of organizational culture. Incident response strategies and disaster recovery plans are crucial for minimizing the impact of cyber incidents when they occur. An effective incident response plan outlines the steps to be taken in the event of a security breach, including identifying the source of the breach, containing the threat, eradicating it, and recovering affected systems. This plan should include clear roles and responsibilities for team members to ensure a coordinated response. Disaster recovery plans complement incident response strategies by focusing on restoring operations after a significant disruption. These plans should include detailed procedures for data recovery, system restoration, and business continuity. Regular testing of these plans is essential to ensure their effectiveness and to identify areas for improvement.

8.6 The Role of Organizations in Cybersecurity

The role of organizations in cybersecurity is increasingly critical as cyber threats evolve and become more sophisticated. An effective cybersecurity strategy is not solely the responsibility of the IT department; it requires a holistic approach that integrates organizational culture, employee training, and collaboration between public and private sectors. Organizational culture plays a significant role in shaping cybersecurity practices within a company. A strong security-first culture encourages employees to prioritize cybersecurity in their daily activities and decision-making processes. When leadership emphasizes the importance of cybersecurity and demonstrates a commitment to protecting sensitive information, it fosters an environment where employees understand their role in safeguarding the organization. This cultural shift can lead to more proactive behaviors, such as reporting suspicious activities and adhering to security protocols. Organizations that successfully cultivate a security-first culture often see a reduction in security incidents and an overall enhancement in their cybersecurity posture. Training and awareness programs for employees are essential components of an effective cybersecurity strategy. These programs educate staff about the various types of cyber threats, such as phishing, malware, and social engineering attacks, and provide them with the knowledge and skills needed to recognize and respond to these threats. Regular training sessions, workshops, and simulations can help employees stay informed about the latest cybersecurity trends and best practices. Moreover, fostering a culture of continuous learning ensures that employees remain vigilant and prepared to address emerging threats. Given that human error is a significant factor in many security breaches, investing in employee training can significantly reduce the risk of successful attacks. Collaboration between public and private sectors is also vital for enhancing cybersecurity efforts. Cyber threats often transcend organizational boundaries, making it essential for different sectors to work together to share information, resources, and best practices. Public agencies, such as the Cybersecurity and Infrastructure Security Agency (CISA), provide valuable guidance and support to organizations in both the public and private sectors. By offering tools, resources, and training, CISA helps organizations implement key cybersecurity practices and strengthen their defenses against cyber threats. Furthermore, partnerships between government and industry can facilitate the sharing of threat intelligence, enabling organizations to stay ahead of potential attacks and better protect their critical assets. In addition to these collaborative efforts, organizations can benefit from participating in information-sharing initiatives and industry-specific cybersecurity forums. These platforms allow organizations to exchange insights on emerging threats, vulnerabilities, and effective mitigation strategies. By learning from the experiences of others, organizations can enhance their own cybersecurity measures and contribute to a more resilient overall security landscape.

8.7 Future Trends in Cybersecurity

As technology continues to advance, the cybersecurity landscape is poised to undergo significant changes in the coming years. By understanding the key trends shaping the future of cybersecurity, organizations can better prepare themselves to combat evolving threats and maintain resilience in the digital age.

One of the most notable trends in the digital threat landscape is the increasing sophistication of cyberattacks. Cybercriminals are continuously developing more advanced techniques to infiltrate systems and steal sensitive data. In the future, we can expect to see a rise in targeted attacks, where threat actors conduct extensive research to identify vulnerabilities and craft personalized strategies for each victim. Additionally, the proliferation of Internet of Things (IoT) devices introduces new attack vectors, as many IoT devices lack robust security measures. As the number of connected devices grows, so does the potential for widespread disruption from large-scale IoT-based attacks.

Artificial intelligence (AI) and machine learning (ML) are poised to play a significant role in shaping the future of cybersecurity. These technologies can enhance threat detection and response by analyzing vast amounts of data to identify patterns and anomalies indicative of cyber threats. AI-driven systems can also predict potential vulnerabilities, recommend proactive security measures, and adapt authentication processes based on user behavior. However, it's important to note that AI can also be leveraged by cybercriminals to automate attacks and craft more convincing phishing schemes. As AI and ML become more prevalent in cybersecurity, organizations must stay vigilant and ensure that their defensive measures keep pace with the evolving threat landscape.

In the face of rapidly changing cyber threats, organizations must adopt a mindset of continuous adaptation and learning. This involves regularly reviewing and updating security policies, conducting risk assessments, and implementing the latest security technologies. Additionally, fostering a culture of cybersecurity awareness and providing ongoing training for employees is crucial for mitigating the risk of human error, which remains a significant factor in many cyber incidents. By embracing a culture of continuous learning and adaptation, organizations can enhance their resilience and better protect their critical assets in the ever-evolving digital landscape. In conclusion, the future of cybersecurity is marked by the increasing sophistication of cyber threats, the growing role of AI and machine learning, and the importance of continuous adaptation and learning. By staying informed about emerging trends and best practices, organizations can proactively address security challenges and maintain a strong cybersecurity posture in the years to come.

8.8 Conclusion

In summary, the landscape of cybersecurity is continually evolving, driven by the increasing sophistication of cyber threats and the rapid advancement of technology. Key points discussed throughout this exploration include the critical role of organizations in fostering a strong security culture, the importance of training and awareness programs for employees, and the necessity of collaboration between public and private sectors to enhance overall security measures. Additionally, the significance of defensive technologies, best practices, and effective incident response strategies cannot be overstated, as they form the foundation of a robust cybersecurity posture. As we look to the future, it is clear that organizations must remain vigilant in the face of emerging threats. The predictions for the evolution of cyber threats indicate a trend toward more targeted and complex attacks, particularly with the rise of IoT devices and the potential misuse of artificial intelligence. To combat these challenges, organizations must leverage AI and machine learning to enhance their security capabilities while also recognizing the dual-use nature of these technologies. The ongoing need for vigilance and innovation in cybersecurity cannot be emphasized enough. Cybersecurity is not a one-time effort but a continuous journey that requires organizations to adapt and evolve in response to new threats and vulnerabilities. Regularly updating security policies, investing in employee training, and embracing a culture of continuous learning are essential components of a proactive cybersecurity strategy. Ultimately, the commitment to cybersecurity must be ingrained in the organizational culture, supported by leadership, and embraced by all employees. By fostering an environment of awareness and preparedness, organizations can significantly reduce their risk of cyber incidents and enhance their resilience against the ever-changing landscape of cyber threats. As we move forward, the combination of vigilance, innovation, and collaboration will be vital in ensuring that organizations can effectively protect their critical assets and maintain the trust of their stakeholders in an increasingly digital world.

References

1. Basholli, F., Juraev, D.A., Egamberdiev, K.: Framework, tools and challenges in cyber security. Karshi Multidiscip. Int. Sci. J. **1**(1), 94–104 (2024)
2. Fidler, B.: Cybersecurity governance: a prehistory and its implications. Digit. Policy Regul. Gov. **19**(6), 449–465 (2017)
3. Yusof, Z.B.: Exploration of advanced persistent threats: techniques, mitigation strategies, and impacts on critical infrastructure. Int. J. Adv. Cybersecur. Syst. Technol. Appl. **8**(12), 1–9 (2024)

4. Wolff, E.D., GroWlEy, K.M., Lerner, M.O., Welling, M.B., Gruden, M.G., Canter, J.: Navigating the solarwinds supply chain attack. Procurement Law. **56**, 3 (2021)
5. Gawazah, L., Rondla, A., Balhareth, M.S.A.: To pay or not to pay: the US colonial pipeline ransomware attack. Thunderbird School of Global Management (2024)
6. Taherdoost, H.: Understanding cybersecurity frameworks and information security standards—a review and comprehensive overview. Electronics. **11**(14), 2181 (2022)
7. Singh, S.: Intrusion detection systems (IDs) and intrusion prevention systems (IPs) For network security: a critical analysis. Adv. J. Wirel. Mobile Commun. **123**, 123–153 (2023)

Chapter 9
3D Printing: From Principles to Diverse Applications

9.1 Introduction

3D printing, also known as additive manufacturing, refers to the process of creating three-dimensional objects from a digital file, typically by layering materials. This technology has revolutionized how products are designed, developed, and manufactured, enabling the creation of complex shapes and structures that would be challenging or impossible to produce using traditional manufacturing methods. The concept of 3D printing dates back to the 1980s, when the first patents were filed, marking the beginning of its evolution. Initially, the technology was limited to industrial applications, but over the decades, it has expanded to various sectors, becoming more accessible and affordable [1].

The historical development of 3D printing has seen significant milestones, from the introduction of Stereolithography (SLA) in 1986, which was the first commercialized 3D printing technology, to the later development of various other techniques such as Fused Deposition Modeling (FDM) and Selective Laser Sintering (SLS). These advancements have broadened the range of materials that can be used in 3D printing and improved the precision and speed of the printing process. The ongoing innovations in 3D printing continue to push the boundaries of what is possible, making it a vital tool in various industries, from healthcare and aerospace to fashion and consumer goods (Fig. 9.1) [2].

The importance of 3D printing in modern manufacturing cannot be overstated. It has transformed traditional manufacturing processes by reducing waste, lowering production costs, and enabling mass customization. The ability to create prototypes rapidly and test them in real-world scenarios has accelerated product development cycles and fostered innovation. Moreover, 3D printing has democratized manufacturing, allowing small businesses and individuals to create and produce goods that were previously only within the reach of large corporations. As 3D printing technology continues to advance, its impact on the global manufacturing landscape will

H. Taherdoost, *Innovating and Integrating Emerging Technologies*, Signals and Communication Technology, https://doi.org/10.1007/978-3-031-93953-2_9

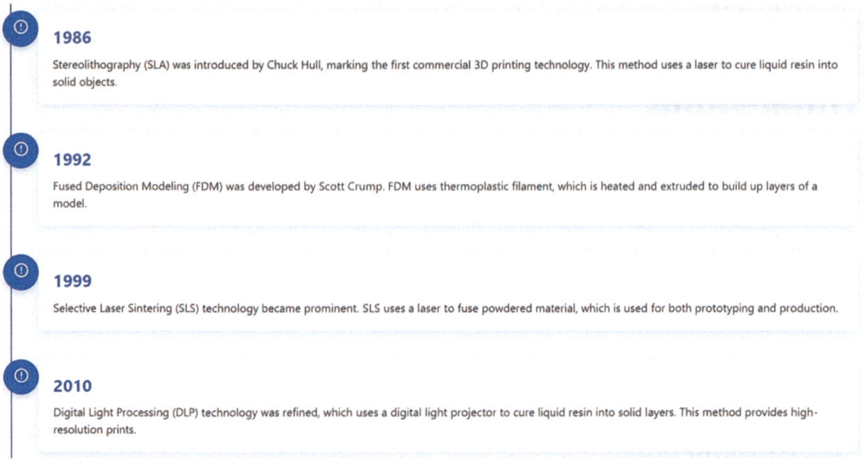

Fig. 9.1 Historical context of 3D printing

likely grow, driving further changes in how products are conceived, developed, and brought to market.

9.2 Principles of 3D Printing

Additive manufacturing, commonly known as 3D printing, represents a revolutionary approach to creating three-dimensional objects from a digital design. This technology builds objects layer by layer, contrasting with traditional subtractive manufacturing methods that involve cutting away material from a larger block. At its core, additive manufacturing enables the creation of complex geometries and structures that would be challenging or impossible to achieve with conventional methods. This process starts with a digital model, typically designed using Computer-Aided Design (CAD) software, which is then translated into a format that the 3D printer can interpret.

One of the fundamental principles behind additive manufacturing is its ability to construct parts by adding material only where it is needed. This contrasts sharply with traditional manufacturing processes that often involve removing excess material from a solid block. By building up material layer by layer, additive manufacturing not only reduces waste but also allows for greater flexibility in design. This technique can accommodate intricate details and complex internal structures that would otherwise be difficult to produce. Furthermore, the ability to create custom, on-demand parts directly from digital designs has streamlined prototyping and product development, making the iterative design process faster and more cost-effective.

The various 3D printing technologies, such as Fused Deposition Modeling (FDM), Stereolithography (SLA), and Selective Laser Sintering (SLS), utilize different methods to achieve the additive manufacturing process. FDM works by extruding a thermoplastic filament through a heated nozzle, which deposits the material in successive layers to build the object. SLA employs a laser to cure a liquid resin, solidifying it layer by layer. SLS, on the other hand, uses a laser to fuse powdered material, creating a solid structure. Each of these methods has its own set of advantages and is suitable for different applications, from rapid prototyping to full-scale production.

Understanding these principles highlights the transformative potential of additive manufacturing across various industries. It not only enables the production of highly customized and intricate parts but also fosters innovation by making it possible to rapidly prototype and iterate on designs. As technology continues to advance, additive manufacturing is poised to play an increasingly important role in shaping the future of production and design.

9.2.1 Key Components of 3D Printing Systems

The key components of 3D printing systems are integral to the functioning of the technology, each playing a specific role in the additive manufacturing process (Fig. 9.2). At the core of a 3D printer is the **printing mechanism**, which is responsible for the actual creation of the object. This mechanism varies depending on the type of 3D printing technology used, such as Fused Deposition Modeling (FDM), Stereolithography (SLA), or Selective Laser Sintering (SLS). The printing mechanism generally includes an **extruder** or **print head** for depositing or solidifying material, and a **build platform** or **print bed** where the object is formed.

Fig. 9.2 Key components of 3D printing systems

The **material delivery system** is another crucial component. This system handles the supply of raw materials to the printing mechanism. In FDM printers, for example, the material is typically in the form of a spool of thermoplastic filament that is fed into the extruder. In SLA and SLS printers, the materials are usually in liquid or powder form and are delivered to the printing area accordingly. The **material properties** and delivery method can greatly influence the quality and characteristics of the final printed object.

Another important component is the **motion control system**, which manages the precise movement of the print head and build platform. This system ensures that the printing mechanism follows the exact paths and layers defined by the digital model. It usually includes motors, belts, and rails, which work together to achieve accurate positioning and movement. The accuracy and speed of the motion control system are critical for achieving high-resolution prints and maintaining consistency across different layers.

The **heating and cooling systems** are also vital, particularly in systems that use thermoplastic materials. For example, in FDM printers, the extruder and print bed are heated to ensure proper melting and adhesion of the filament. Conversely, in SLA printers, the cooling system helps manage the curing process of the resin. Proper temperature regulation is essential for maintaining material flow and preventing defects in the printed object.

Finally, the **software and user interface** are essential for operating the 3D printing system. The software includes slicing programs that convert digital 3D models into instructions for the printer. This software is responsible for generating the paths and layers that the printer will follow. The user interface, which can be a physical control panel or a software application, allows users to configure settings, start and stop prints, and monitor the progress of the printing process. This component ensures that users can interact effectively with the printer and manage print jobs with precision.

Each of these components plays a specific role in the 3D printing process, contributing to the overall functionality and effectiveness of the system. Understanding these key elements is crucial for optimizing the performance and capabilities of 3D printing technology.

9.2.2 Overview of 3D Printing Processes

Fused Deposition Modeling (FDM) is a widely used 3D printing technology that involves extruding a thermoplastic filament through a heated nozzle. This process builds objects layer by layer, with each layer adhering to the one below it as the filament cools and solidifies. The versatility and cost-effectiveness of FDM make it a popular choice for both prototyping and production of functional parts.

Stereolithography (SLA) utilizes a different approach by employing a laser to cure liquid resin in a vat. The laser traces out the cross-sections of the object, solidifying the resin layer by layer. SLA is known for its high precision and smooth surface finishes, which are particularly valuable for applications requiring detailed and accurate models.

Selective Laser Sintering (SLS) involves using a laser to selectively fuse powdered material into solid structures. The laser scans the surface of the powder bed and sinters particles together to form each layer. SLS supports a wide range of materials and produces parts with good mechanical properties and complex geometries, making it suitable for functional prototypes and end-use parts.

Digital Light Processing (DLP) is similar to SLA but uses a digital light projector to cure the resin instead of a laser. This method allows for faster printing speeds as it can cure an entire layer at once. DLP also provides high resolution and precision, making it ideal for applications requiring fine details and smooth finishes.

Binder jetting operates by depositing a binder material onto a bed of powder to selectively bind particles together. After each layer of binder is applied, a new layer of powder is spread over the previous one. The process is repeated until the object is complete. Binder jetting is capable of producing large, complex parts and is used for applications ranging from industrial parts to artistic creations.

Emerging technologies in 3D printing include several innovative approaches that are still in development or gaining traction. These technologies aim to improve printing speed, material capabilities, and application versatility. Examples include multi jet fusion (MJF), which uses multiple agents to fuse powder, and continuous liquid interface production (CLIP), which enables rapid printing by continuously growing objects in a pool of liquid resin. These advancements are expanding the boundaries of what 3D printing can achieve and opening new possibilities for various industries. Table 9.1 provides a clear and organized overview of each 3D printing technology, highlighting their unique characteristics, benefits, and typical applications.

9.3 Materials Used in 3D Printing

In 3D printing, selecting the right material is essential for achieving desired properties and performance in the final printed object. The choice of material affects everything from the functionality to the cost of the printed part. This overview explores the range of materials used in additive manufacturing, highlighting their applications and benefits (Fig. 9.3).

Plastics and polymers are among the most commonly used materials in 3D printing due to their versatility and affordability. In Fused Deposition Modeling (FDM), which is one of the most widely used 3D printing technologies, thermoplastics are the primary materials. Thermoplastics, such as polylactic acid (PLA), acrylonitrile butadiene styrene (ABS), and polyethylene terephthalate glycol (PETG), are known for their ease of use and adaptability. PLA is popular for its user-friendliness and environmental benefits, as it is biodegradable and derived from renewable resources. ABS, on the other hand, offers strength and durability, making it suitable for functional prototypes and parts that require impact resistance. PETG combines desirable

Table 9.1 Overview of 3D printing technologies

Technology	Description	Material	Advantages	Applications
Fused deposition modeling (FDM)	Extrudes thermoplastic filament through a heated nozzle, building objects layer by layer	Thermoplastics (e.g., PLA, ABS)	Cost-effective, versatile, widely available	Prototyping, functional parts, hobbyist projects
Stereolithography (SLA)	Uses a laser to cure liquid resin layer by layer in a vat	Liquid resin	High precision, smooth surface finishes	Detailed models, jewelry, dental applications
Selective laser sintering (SLS)	Fuses powdered material using a laser, layer by layer.	Powders (e.g., nylon, metal)	Supports complex geometries, good mechanical properties	Functional prototypes, end-use parts
Digital Light Processing (DLP)	Cures resin layer by layer using a digital light projector	Liquid resin	Fast printing speeds, high resolution	Detailed models, rapid prototyping
Binder jetting	Deposits binder onto a powder bed, layer by layer, to bind particles together	Powders (e.g., metal, sand)	Capable of producing large and complex parts	Industrial parts, artistic creations
Multi jet fusion (MJF)	Uses multiple agents to fuse powder in a single pass	Powders (e.g., nylon)	High speed, good material properties	Functional parts, prototypes
Continuous liquid interface production (CLIP)	Uses a continuously flowing resin pool to print objects rapidly	Liquid resin	Rapid printing, high resolution	Prototyping, custom parts

attributes from both PLA and ABS, including strength, flexibility, and resistance to chemicals, making it a versatile choice for a wide range of applications.

Stereolithography (SLA) employs photopolymer resins, which are cured using ultraviolet (UV) light. These resins are available in several formulations to meet different needs. Standard resins are used for high-detail and precise prints, while tough resins are engineered to withstand impact and stress, making them ideal for parts that need to endure mechanical forces. Flexible resins are designed to provide elasticity and are used in applications requiring parts that can bend or stretch. High-temperature resins are capable of withstanding elevated temperatures and are suited for applications where thermal resistance is crucial.

In the realm of metal 3D printing, various metals and alloys are employed to produce high-strength and functional components. Technologies such as Selective Laser Sintering (SLS) and Direct Metal Laser Sintering (DMLS) are commonly

Fig. 9.3 Overview of 3D printing materials

used for metal additive manufacturing. These technologies use lasers to fuse metal powders into solid structures, allowing for the production of parts with complex geometries and exceptional mechanical properties. Stainless steel, titanium, and aluminum are frequently used metals due to their strength, corrosion resistance, and versatility. Stainless steel is valued for its robustness and resistance to oxidation, making it suitable for functional parts and industrial applications. Titanium is known for its high strength-to-weight ratio and biocompatibility, which makes it ideal for aerospace and medical implants. Aluminum is appreciated for its lightweight and machinability, often used in automotive and aerospace industries.

Ceramics and composites are another important category of 3D printing materials, offering unique properties and applications. Ceramic 3D printing can produce parts with high-temperature resistance and excellent hardness. Ceramics are commonly used in applications such as dental implants, prosthetics, and high-performance industrial components. The additive manufacturing process for ceramics involves printing with a paste or slurry that is later fired in a kiln to achieve the desired hardness and strength.

Composites combine different materials to enhance performance and create parts with tailored properties. Carbon fiber-reinforced polymers are a notable example, where carbon fibers are embedded within a polymer matrix to improve strength, rigidity, and lightweight characteristics. This makes composite materials suitable for high-performance applications, such as in aerospace and automotive industries. Other composites include glass fiber-reinforced materials and metal matrix composites, each offering specific advantages in terms of strength, durability, and resistance to environmental factors.

Bio-based materials are an emerging area in 3D printing, reflecting a growing interest in sustainability and environmentally friendly practices. These materials include biodegradable polymers and bio-resins derived from natural sources. For instance, bio-based PLA is made from renewable plant resources, which reduces the environmental impact compared to traditional petroleum-based plastics. Other examples include algae-based resins and proteins, which are being explored for their potential to offer sustainable alternatives to conventional materials.

Recent innovations in 3D printing materials continue to expand the capabilities and applications of additive manufacturing. Advances include the development of high-performance thermoplastics with enhanced thermal, mechanical, and chemical properties, as well as the integration of functional materials such as conductive inks and self-healing polymers. These innovations enable the creation of complex, multifunctional parts and open new possibilities for industries ranging from electronics to healthcare.

The diverse range of materials available for 3D printing offers significant flexibility in designing and producing parts for various applications. By understanding the properties and benefits of different materials, manufacturers can select the most suitable option for their specific needs, whether for rapid prototyping, functional components, or high-performance applications.

9.4 Design and Modeling for 3D Printing

Design and modeling are crucial aspects of 3D printing, influencing the accuracy, functionality, and overall success of printed objects. The process begins with computer-aided design (CAD), where designers use specialized software to create detailed 3D models. CAD software allows for the precise manipulation of geometric shapes, enabling the design of complex structures that can be transformed into physical objects through 3D printing. This step is fundamental as it provides the digital blueprint from which the printer will work.

File formats and standards play a significant role in ensuring that CAD models are correctly interpreted by 3D printers. Common file formats such as STL (Stereolithography), OBJ (Wavefront Object), and AMF (Additive Manufacturing File) each have their specific uses and characteristics. STL is widely used due to its simplicity and compatibility with most 3D printers, providing a basic representation of a model's geometry. OBJ supports more detailed information, including texture and color, making it suitable for more complex applications. AMF, an XML-based format, offers advanced features such as color and material definitions, catering to more sophisticated printing needs [3].

Design considerations and constraints are pivotal in 3D printing. Designers must account for various factors such as material properties, print resolution, and structural integrity. For instance, certain designs may require support structures during printing to prevent collapsing or deformation. Additionally, the orientation of the model and the choice of infill density can impact the strength and appearance of the final product. Understanding these constraints helps in optimizing the design for both functionality and printability.

Software tools and workflow optimization are essential for streamlining the design and printing process. Numerous software tools are available that assist in various stages of 3D printing, from initial design to final production. These tools often include features for model repair, slicing (the process of converting a 3D model into printable layers), and simulation to predict print outcomes. Workflow optimization involves integrating these tools effectively to reduce time and costs while improving the quality and efficiency of the printing process. By leveraging advanced software and optimizing workflows, designers can ensure that their models are accurately translated into high-quality printed objects.

9.5 Applications of 3D Printing

3D printing has transformed numerous industries through its innovative approach to manufacturing and design (Fig. 9.4). In the aerospace and automotive sectors, 3D printing has become an integral technology for producing complex components that are both lightweight and durable. Aerospace companies utilize 3D printing to create intricate parts that reduce the weight of aircraft, which leads to improved fuel efficiency and performance. This technology allows for the rapid prototyping of parts, speeding up the development process and enabling more customized solutions. In the automotive industry, 3D printing facilitates the production of lightweight components, custom parts, and prototypes, accelerating innovation and reducing production costs.

The healthcare and biomedical fields have also seen significant advancements due to 3D printing. One of the notable applications is in the creation of prosthetics

Fig. 9.4 3D printing applications across industries

and implants. Custom prosthetics, tailored to the individual's unique anatomy, can be produced with greater precision and at a lower cost compared to traditional methods. Implants, such as those used in orthopedic surgery, benefit from 3D printing by offering better fit and functionality. Beyond prosthetics and implants, bioprinting and tissue engineering represent exciting frontiers in healthcare. Bioprinting involves creating tissue-like structures using bio-inks composed of living cells, which can potentially lead to breakthroughs in organ replacement and regenerative medicine. Tissue engineering focuses on developing functional tissues and organs, aiming to address shortages and improve patient outcomes.

In architecture and construction, 3D printing is revolutionizing the way buildings and structures are designed and constructed. This technology allows for the creation of complex architectural features that were previously difficult or impossible to achieve using traditional methods. Large-scale 3D printers are capable of constructing entire buildings or significant portions of them, which can reduce construction time and costs. Additionally, 3D printing enables the use of innovative materials and techniques, potentially leading to more sustainable and efficient construction practices.

Consumer goods and electronics have also benefited from 3D printing, particularly in the customization and rapid production of products. In the consumer goods sector, 3D printing allows for the creation of bespoke items tailored to individual preferences, such as personalized phone cases or custom-designed jewelry. Electronics manufacturers use 3D printing to produce prototype components, enabling quicker iteration and testing of new designs. This rapid prototyping capability accelerates the development of new technologies and products, keeping pace with the fast-moving electronics market.

Fashion and art have embraced 3D printing as a means to push creative boundaries and explore new design possibilities. Designers in the fashion industry use 3D printing to create unique garments, accessories, and footwear that combine traditional techniques with innovative manufacturing methods. The ability to design and print complex patterns and structures opens up new avenues for artistic expression and customization. Artists leverage 3D printing to produce intricate sculptures and installations, allowing for the realization of designs that would be challenging to achieve with conventional techniques.

Education and research benefit from 3D printing through its role in enhancing learning and experimentation. In educational settings, 3D printers provide students with hands-on experience in design, engineering, and manufacturing. They offer a practical way to understand complex concepts and create tangible models of theoretical ideas. In research, 3D printing facilitates the rapid prototyping of experimental devices and components, enabling researchers to test and refine their innovations more efficiently.

The food industry and culinary arts are exploring the potential of 3D printing to revolutionize food production and presentation. 3D food printers can create intricate and customized food designs, ranging from personalized chocolate shapes to complex sugar structures. This technology allows chefs and food designers to experiment with new textures and flavors, potentially transforming the culinary experience. Additionally, 3D printing in food production offers the possibility of more sustainable practices by optimizing ingredient usage and reducing waste.

9.6 Challenges and Limitations of 3D Printing

The field of 3D printing, while revolutionary, faces several challenges and limitations that affect its widespread adoption and application. These challenges can be categorized into technical challenges, material limitations, economic considerations, and regulatory and ethical issues.

Technical challenges in 3D printing primarily revolve around the complexity and precision required in the printing process. One significant issue is the resolution and accuracy of 3D printers, which can impact the final quality of printed objects. Many printers struggle to achieve the high precision needed for intricate designs, which can result in defects or a lack of detail in the final product. Additionally, the speed of printing is often a constraint; while prototyping can be quick, producing high-quality, functional parts can be time-consuming. This limitation can hinder the efficiency of manufacturing processes, especially in industries that require rapid production.

Material limitations are another critical challenge. Although the range of materials available for 3D printing has expanded, many of these materials have constraints that affect their usability. For example, some materials may not possess the required strength, flexibility, or durability for specific applications. The lack of high-performance materials suitable for certain industries, such as aerospace or medical implants, limits the effectiveness of 3D printing in these areas. Furthermore, the cost and availability of these materials can be prohibitive, particularly for small-scale operations or individual users.

Economic considerations also play a significant role in the adoption of 3D printing technologies. The initial investment in high-quality 3D printers and related equipment can be substantial. While costs have been decreasing, advanced printers that offer high precision and speed remain expensive. Additionally, the ongoing costs of maintenance, materials, and post-processing can add to the overall expense. For businesses, especially smaller ones, these costs can be a barrier to entry, limiting the widespread adoption of 3D printing technologies.

Regulatory and ethical issues present complex challenges for the integration of 3D printing into various industries. As the technology evolves, there is a need for clear guidelines and standards to ensure the safety and quality of printed products. For instance, in the medical field, ensuring that 3D-printed implants and prosthetics meet rigorous safety standards is crucial. Moreover, intellectual property concerns arise as 3D printing makes it easier to reproduce and distribute designs, potentially leading to disputes over copyright and patent rights. Ethical considerations, such as the potential for misuse in creating illegal items or the environmental impact of certain materials, also need to be addressed. Establishing comprehensive regulations and ethical frameworks is essential for the responsible advancement of 3D printing technology.

9.7 Future Trends in 3D Printing

The future of 3D printing is poised for significant transformation, driven by advancements in technology and innovative applications across various sectors. As we look ahead, several key trends are expected to shape the evolution of 3D printing, from technological advancements to its impact on global supply chains and sustainability.

Advancements in 3D printing technologies are set to revolutionize the industry further. Emerging techniques and improvements in existing technologies promise to enhance the precision, speed, and capabilities of 3D printers. New materials with better performance characteristics are being developed, expanding the range of possible applications. Innovations such as multi-material printing and advanced composite materials will enable the creation of more complex and functional objects. Additionally, improvements in printer hardware, including faster processors and more precise control systems, are expected to push the boundaries of what 3D printing can achieve.

The integration of artificial intelligence (AI) and machine learning into 3D printing processes is another exciting development. AI can optimize printing parameters, predict and correct errors, and improve the overall efficiency of the manufacturing process. Machine learning algorithms can analyze vast amounts of data to refine design and production techniques, leading to higher-quality outputs and reduced waste. AI-driven design tools are also emerging, allowing for the creation of more intricate and optimized structures that were previously difficult to achieve.

Sustainability and green practices are becoming increasingly important in the realm of 3D printing. The industry is exploring ways to minimize environmental impact through the development of eco-friendly materials and processes. This includes the use of recycled materials, biodegradable polymers, and energy-efficient printing techniques. The concept of circular economy is gaining traction, where 3D printing can contribute to reducing waste by enabling on-demand production and reducing the need for large inventories. As awareness of environmental issues grows, the push for sustainable practices in 3D printing will likely accelerate.

The impact of 3D printing on global supply chains is substantial. The technology has the potential to disrupt traditional manufacturing and distribution models by enabling localized production and reducing the reliance on centralized factories. This shift could lead to more agile and responsive supply chains, with products being manufactured closer to their point of use. On-demand production can also reduce the need for large-scale warehousing and transportation, leading to cost savings and reduced carbon footprints. The ability to rapidly produce customized products locally can further enhance supply chain efficiency and flexibility.

The future of mass customization is closely linked to advancements in 3D printing technology. As printers become more capable and accessible, the potential for personalized and bespoke products will expand. Consumers will increasingly demand customized solutions tailored to their individual preferences, whether in fashion, healthcare, or consumer goods. 3D printing enables the efficient production of small batches of unique items, allowing businesses to offer personalized products at scale. This trend toward mass customization will drive innovation in product design and manufacturing, offering new opportunities for businesses to differentiate themselves in the market.

9.8 Conclusion

The exploration of 3D printing, from its foundational principles to its diverse applications, reveals both its transformative potential and its current limitations. As we summarize the key concepts, the evolving landscape, and the final reflections on this technology, it becomes clear that 3D printing stands at the forefront of a new era in manufacturing and design.

The key concepts of 3D printing encompass the technology's ability to create three-dimensional objects through additive processes. By building up materials layer by layer, 3D printing offers unparalleled design flexibility and the ability to produce complex geometries that traditional manufacturing methods struggle to achieve. The technology leverages various materials, from plastics to metals, each contributing to different applications across industries. The principles of 3D printing, including the types of printing technologies such as Fused Deposition Modeling (FDM), Stereolithography (SLA), and Selective Laser Sintering (SLS), provide a foundation for understanding how this technology operates and its various use cases.

As the landscape of 3D printing continues to evolve, several key trends are shaping its future. Technological advancements are driving improvements in speed, accuracy, and material diversity, enabling more complex and functional products. The integration of artificial intelligence and machine learning is optimizing printing processes and enhancing design capabilities. Sustainability is becoming a focal point, with efforts to develop eco-friendly materials and reduce waste. The impact of 3D printing on global supply chains is also significant, offering the potential for localized production and reducing dependency on traditional manufacturing and distribution models. The trend toward mass customization allows for personalized products at scale, further illustrating the versatility and innovation inherent in 3D printing technology.

Reflecting on the progress and potential of 3D printing, it is evident that the technology is poised to revolutionize various sectors, from healthcare and aerospace to consumer goods and education. The ability to rapidly prototype, customize, and produce complex designs offers opportunities for enhanced creativity and efficiency. However, challenges remain, including technical limitations, material constraints, and the need for regulatory frameworks to ensure safety and ethical use. As 3D printing continues to advance, addressing these challenges will be crucial in realizing its full potential.

In conclusion, 3D printing represents a dynamic and rapidly evolving field with the capacity to transform how we design, manufacture, and consume products. Its ongoing development promises to unlock new possibilities and drive innovation across multiple domains. As we move forward, embracing the opportunities and addressing the challenges will be key to harnessing the full benefits of this groundbreaking technology.

References

1. Balletti, C., Ballarin, M., Guerra, F.: 3D printing: state of the art and future perspectives. J. Cult. Herit. **26**, 172–182 (2017)
2. Robles Martinez, P., Basit, A.W., Gaisford, S.: The history, developments and opportunities of stereolithography. In: 3D Printing of Pharmaceuticals, pp. 55–79. Springer, Cham (2018)
3. Kumar, A., Kumar, P., Mittal, R.K., Singh, H.: Printing file formats for additive manufacturing technologies. In: Advances in Additive Manufacturing, pp. 87–102. Elsevier, Amsterdam (2023)

Chapter 10
Empowering Ideas and Ventures with Crowdfunding

10.1 Introduction

Crowdfunding is a method of raising capital by gathering small amounts of money from a large number of people, typically through online platforms. This approach democratizes access to funding, allowing individuals and businesses to secure financial support without relying on traditional sources such as banks or venture capitalists. The concept of crowdfunding has evolved over time, originating from community-based fundraising efforts and gradually moving into the digital age with the advent of online platforms (Fig. 10.1). Early forms of crowdfunding can be traced back to the eighteenth century when public subscriptions helped fund projects like the construction of the Statue of Liberty's pedestal. However, the modern era of crowdfunding began in the early 2000s with the launch of platforms like Kickstarter and Indiegogo, which made it easier for creators to reach a global audience and raise funds through the internet. Since then, crowdfunding has continued to grow and diversify, encompassing various models such as donation-based, reward-based, equity-based, and debt-based crowdfunding, each catering to different types of projects and funding needs [1].

10.2 Types of Crowdfunding

Crowdfunding has evolved into a multifaceted approach for raising capital, each type catering to distinct needs and goals (Fig. 10.2). Donation-based crowdfunding involves raising funds for a cause, project, or individual without expecting any financial return. Donors contribute out of goodwill or to support a cause they believe in, and they do not receive any monetary compensation or equity in return. This type

H. Taherdoost, *Innovating and Integrating Emerging Technologies*, Signals and Communication Technology, https://doi.org/10.1007/978-3-031-93953-2_10

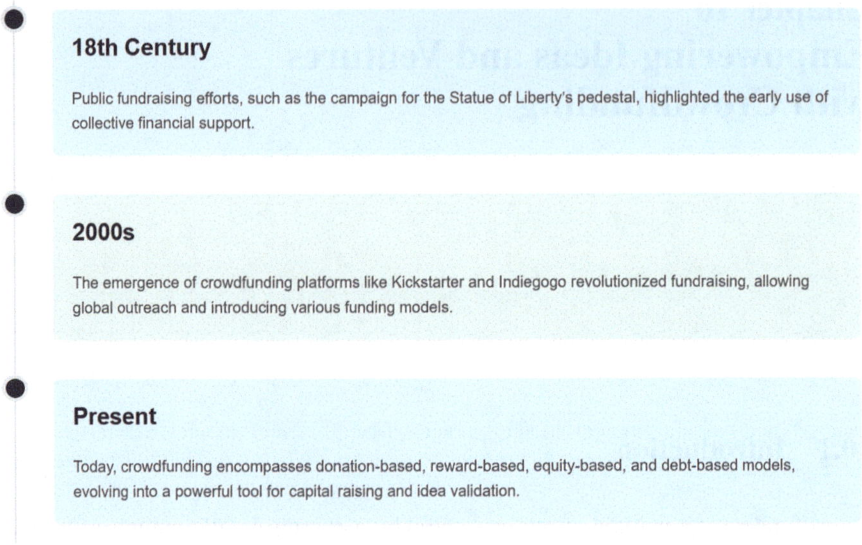

Fig. 10.1 Historical timeline of crowdfunding

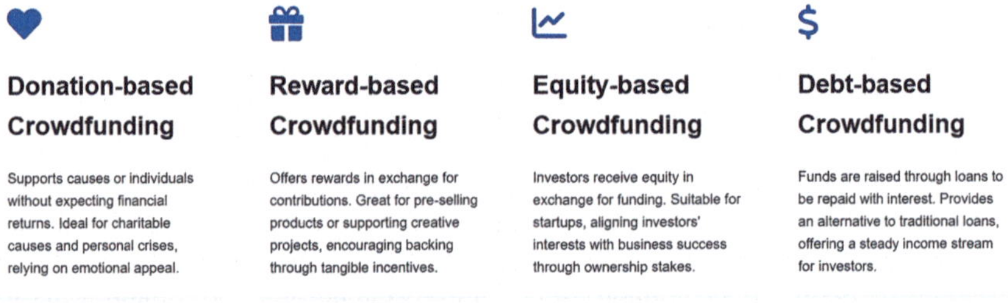

Fig. 10.2 Types of crowdfunding

is often used for charitable causes, personal emergencies, community projects, or non-profit initiatives. The success of such campaigns typically relies on a compelling story or emotional appeal that engages potential donors. The simplicity of donation-based crowdfunding makes it easy to set up and manage, with minimal legal and financial complexities, and it can attract contributions from a broad audience interested in supporting a cause.

In contrast, reward-based crowdfunding involves raising funds by offering non-financial incentives or rewards to backers. These rewards are typically tied to the project or product being funded and vary based on the contribution level. Contributors receive tangible rewards such as products, services, or experiences instead of financial returns or equity. Reward-based crowdfunding often includes tiered contribution levels, where different amounts of funding correspond to various rewards. This model is commonly used for pre-selling products or services before they are manufactured or launched. It is particularly useful for market validation, as it helps gauge interest and secure pre-orders. The promise of rewards motivates supporters to contribute, making it an attractive option for launching new consumer products, creative projects, or entrepreneurial ventures.

Equity-based crowdfunding, on the other hand, allows investors to contribute funds in exchange for ownership shares or equity in the business. This type of crowdfunding offers investors a stake in the company, potentially yielding financial returns if the business succeeds. Equity-based crowdfunding is particularly suited for startups and early-stage companies seeking investment to grow and scale their operations. It provides a way for entrepreneurs to raise capital from a larger pool of investors who are interested in becoming part owners of the business. The exchange of equity aligns the interests of investors with the success of the company, making it a powerful tool for funding ventures with high growth potential.

Lastly, debt-based crowdfunding, also known as peer-to-peer lending or crowdlending, involves raising funds through loans that must be repaid with interest. This type provides an opportunity for individuals or businesses to borrow money from a large number of lenders, who receive regular interest payments and the return of their principal investment over time. Debt-based crowdfunding is suitable for projects or businesses seeking capital without giving away equity. It offers an alternative to traditional bank loans and can be appealing to investors looking for a steady income stream from interest payments. The terms of the loan, including interest rates and repayment schedules, are typically outlined at the start of the campaign.

Each type of crowdfunding serves different purposes and offers distinct benefits, making it essential for project creators and investors to choose the model that best fits their needs and goals.

10.3 The Rise of Crowdfunding Platforms

The rise of crowdfunding platforms represents a significant shift in how individuals and businesses can raise capital, seek support, and validate ideas (Table 10.1). This evolution has democratized the funding landscape, allowing a diverse range of projects to access financial backing from a broad audience. Each platform caters to specific types of projects and funding needs, making it essential for project creators to select the one that best aligns with their objectives.

Table 10.1 Crowdfunding platforms comparison

Feature	Kickstarter	Indiegogo	GoFundMe	SeedInvest
Creative projects	✓	✓	✗	✗
Flexible funding	✗	✓	✗	✗
Personal and charitable fundraising	✗	✗	✓	✗
Equity investment	✗	✗	✗	✓

Kickstarter is one of the most well-known crowdfunding platforms, primarily focused on creative and innovative projects. Launched in 2009, Kickstarter has become synonymous with crowdfunding, supporting a wide array of projects from tech gadgets and games to art and film. The platform operates on an all-or-nothing model, meaning that if a project does not reach its funding goal, no funds are collected. This model creates a sense of urgency and encourages backers to contribute early to ensure the project's success. Kickstarter's emphasis on creativity and innovation has helped many projects gain visibility and achieve their funding goals [2].

Indiegogo, founded in 2008, is another prominent player in the crowdfunding space. Unlike Kickstarter, Indiegogo offers both flexible and fixed funding options. This flexibility allows creators to keep funds even if they do not reach their full goal, providing a safety net for those who may not meet their target. Indiegogo supports a broad spectrum of projects, including technology, health, and social causes. Its global reach and varied funding models make it a versatile choice for creators looking to launch their projects internationally [3].

GoFundMe, established in 2010, focuses primarily on personal and charitable fundraising. It is widely recognized for its role in helping individuals raise money for personal emergencies, medical expenses, and community-driven causes. Unlike Kickstarter and Indiegogo, GoFundMe does not operate on a reward or equity basis but instead facilitates direct donations. This platform has been instrumental in enabling people to seek financial support for personal or philanthropic reasons, often through emotionally compelling campaigns [4].

SeedInvest is a leading equity crowdfunding platform, catering to startups and early-stage companies seeking investment. Launched in 2012, SeedInvest provides a platform for accredited investors to invest in high-potential startups in exchange for equity. This model allows entrepreneurs to raise capital from a pool of investors while offering them ownership stakes in the company. SeedInvest's focus on equity-based crowdfunding is particularly valuable for startups that need substantial funding and are willing to share ownership with investors [5].

Each of these platforms has played a pivotal role in the evolution of crowdfunding, offering unique features and catering to different types of funding needs. Kickstarter's focus on creative projects, Indiegogo's flexible funding options, GoFundMe's personal and charitable fundraising, and SeedInvest's equity investment opportunities illustrate the diverse applications of crowdfunding. As the industry continues to grow, these platforms will likely adapt and innovate to meet the changing needs of project creators and investors alike.

10.4 The Benefits of Crowdfunding for Ventures

Crowdfunding has transformed the way individuals and businesses raise capital, offering a range of advantages that go beyond just financial support (Fig. 10.3). One of the most significant benefits of crowdfunding is the democratization of capital access. Traditionally, securing funding often involved navigating complex networks

Fig. 10.3 Benefits of crowdfunding

of investors, banks, or venture capitalists, which could be daunting for startups and small businesses. Crowdfunding platforms, however, enable ventures to reach out to a vast pool of potential backers from around the world. This method allows creators to solicit contributions directly from individuals interested in their ideas, bypassing traditional financial institutions.

Moreover, crowdfunding provides valuable market validation. Before committing to extensive production or development, entrepreneurs can test their ideas by gauging the level of interest from potential customers. A successful crowdfunding campaign can serve as an indicator that there is a demand for the product or service, helping to minimize the risk of investing in unproven concepts. This feedback loop allows for adjustments and improvements based on real-world responses, which can be crucial for refining the product and increasing its chances of success.

Building a community is another key advantage of crowdfunding. Engaging with early backers through a campaign helps to foster a dedicated supporter base. These early adopters are not just financial contributors; they often become passionate advocates for the venture. Their support can extend beyond the initial campaign, as they share their enthusiasm with their networks, provide valuable feedback, and help build momentum for the project. This sense of community can lead to long-term customer loyalty and advocacy, which are invaluable for the ongoing success of the venture.

Crowdfunding campaigns also serve as effective marketing tools. A well-crafted campaign can generate significant exposure and raise awareness about the venture. By leveraging the campaign's platform, entrepreneurs can attract media coverage, engage with potential customers, and create buzz around their project. This increased visibility can lead to additional opportunities, such as partnerships or collaborations, and can help build a stronger brand presence.

10.5 Crafting a Successful Crowdfunding Campaign

When crafting a successful crowdfunding campaign, the first step is to define clear and achievable goals (Fig. 10.4). This involves setting specific funding targets that are both realistic and compelling to potential backers. It is important to be

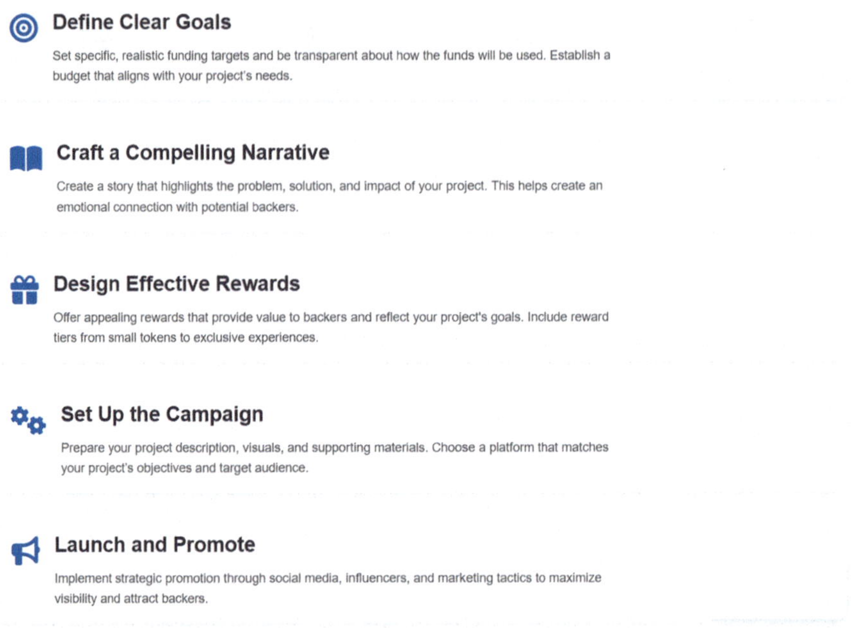

Fig. 10.4 Steps to craft a successful crowdfunding campaign

transparent about how the funds will be used and to establish a budget that aligns with the project's needs.

Developing a compelling narrative is crucial in capturing the interest of potential backers. This narrative should tell a story that resonates with the audience, highlighting the problem being addressed, the solution offered, and the impact of their support. A well-crafted story helps create an emotional connection and motivates people to contribute to the campaign.

Designing effective rewards and incentives is another key aspect. Offering appealing reward tiers can significantly enhance the attractiveness of the campaign. Rewards should be thoughtfully designed to provide value to backers while also reflecting the project's goals. The rewards can range from small tokens of appreciation to exclusive products or experiences, depending on the level of contribution.

Setting up the campaign involves choosing the right crowdfunding platform and preparing all necessary campaign materials. This includes creating a detailed project description, high-quality visuals, and any supporting documents or videos. Selecting a platform that aligns with the project's objectives and target audience is essential for a successful launch.

Finally, launching and promoting the campaign require strategic planning to maximize visibility and engagement. This includes leveraging social media, reaching out to influencers, and employing various marketing tactics to generate buzz. Effective promotion ensures that the campaign reaches a wider audience and attracts potential backers, thereby increasing the likelihood of achieving the funding goal.

10.6 Managing Your Crowdfunding Campaign

Managing a crowdfunding campaign involves several critical aspects to ensure its success. Effective communication with backers is essential; it is important to keep them informed and engaged throughout the campaign. Regular updates, responding to queries, and sharing milestones can help build trust and maintain enthusiasm among supporters.

Handling challenges and making adjustments is another crucial part of campaign management. Setbacks may occur, and it is necessary to navigate these issues calmly and make any necessary changes to the campaign strategy. Addressing problems proactively can help keep the campaign on track and maintain backer confidence.

Tracking progress and celebrating achievements are also key elements. Monitoring campaign milestones and acknowledging when goals are reached not only keeps the campaign on course but also boosts morale and encourages continued support. Celebrating these milestones can further engage backers and enhance the overall success of the campaign.

10.7 Managing Your Crowdfunding Campaign

Once a crowdfunding campaign concludes, effectively managing the funds and resources obtained is critical for ensuring the project's continued success. This involves careful allocation and utilization of the funds to meet the promises made during the campaign. Proper financial management is essential for executing the project as planned, covering production costs, and managing any unexpected expenses. Transparent and efficient use of the funds builds credibility and trust with backers, who have invested their money with the expectation that it will be used responsibly.

Delivering on the promises made during the campaign is another crucial step. This includes ensuring the timely delivery of rewards and keeping backers updated with progress reports. Meeting deadlines and maintaining open communication are vital for upholding the project's reputation and satisfying backer expectations. Providing regular updates helps manage expectations and builds ongoing support, which can be important for future endeavors.

Transitioning to growth after a successful crowdfunding campaign involves leveraging the momentum gained to further develop and expand the project. The success of the campaign can serve as a springboard for additional opportunities, such as scaling operations, enhancing product offerings, or pursuing new markets. Strategic planning and continued engagement with the supporter base can help capitalize on the success, turning initial funding into long-term growth and sustainability. By building on the foundation laid during the campaign, ventures can achieve greater milestones and foster lasting relationships with their backers.

10.8 Future Trends in Crowdfunding

The future of crowdfunding is being significantly influenced by emerging technologies and innovations. Advancements in technology are reshaping how crowdfunding platforms operate, making the process more efficient and accessible. New tools and technologies, such as blockchain and artificial intelligence, are being integrated into crowdfunding platforms to enhance transparency, security, and personalization. Blockchain, for instance, provides a decentralized ledger that can improve trust and reduce fraud, while AI can help tailor campaigns to target specific audiences more effectively. These technological advancements are likely to make crowdfunding more secure and user-friendly, opening up new possibilities for both campaign creators and backers.

Another critical aspect of the future of crowdfunding is the evolving regulatory landscape. As the industry grows, governments and regulatory bodies are increasingly scrutinizing crowdfunding practices. Potential changes in regulations could impact how crowdfunding campaigns are conducted and managed. Regulations may become more stringent to protect investors and ensure transparency, which could affect the way campaigns are structured and promoted. Understanding and adapting to these regulatory changes will be essential for campaign creators to navigate the legal complexities and maintain compliance.

Globally, crowdfunding practices are varied and evolving. Different regions have unique approaches and trends that reflect local economic, cultural, and regulatory environments. For instance, while crowdfunding has seen rapid growth in the United States and Europe, other regions such as Asia and Africa are also experiencing significant development. Each region presents its own set of opportunities and challenges, influenced by local market conditions and technological adoption. As crowdfunding continues to expand globally, understanding these regional differences will be crucial for campaign creators aiming to reach an international audience and leverage global crowdfunding trends.

10.9 Conclusion

In conclusion, the chapter provides a comprehensive overview of the transformative potential of crowdfunding for empowering ideas and ventures. It highlights several key insights, including the democratization of funding, the ability to validate market interest, and the opportunity to build a supportive community around a project. Crowdfunding not only offers an alternative source of capital but also serves as a valuable marketing tool and a platform for early engagement with potential customers.

The discussion emphasizes the importance of defining clear goals, crafting compelling narratives, and designing effective rewards to create a successful crowdfunding campaign. It also underscores the significance of managing funds responsibly and maintaining transparent communication with backers throughout the campaign. By leveraging the momentum gained from a successful crowdfunding effort, ventures can transition into growth phases, explore new opportunities, and build long-lasting relationships with their supporters.

As readers reflect on these insights, they are encouraged to take action and consider how crowdfunding could benefit their own projects and ideas. Whether launching a new product, funding a creative endeavor, or testing a business concept, crowdfunding offers a powerful platform to bring visions to life and achieve meaningful outcomes. By understanding and applying the principles discussed, individuals and ventures can harness the potential of crowdfunding to drive innovation and achieve their goals.

References

1. Bashir, I., Banze, R.: Crowdfunding: The Story of People. Project Nile, Mumbai (2020)
2. Forbes, H., Schaefer, D.: Guidelines for successful crowdfunding. Proc. CIRP. **60**, 398–403 (2017)
3. Parhankangas, A., Mason, C., Landström, H.: Crowdfunding: an introduction. In: Handbook of Research on Crowdfunding, pp. 1–21. Edward Elgar Publishing, Cheltenham (2019)
4. Wade, M.: 'The giving layer of the internet': a critical history of GoFundMe's reputation management, platform governance, and communication strategies in capturing peer-to-peer and charitable giving markets. J. Philanthr. Mark. **28**(4), e1777 (2023)
5. Adwani, R.: Innovative Financing Models for Startups: Challenges and Opportunities in a Globalized Economy. Available at SSRN 5162026 (2025)

Chapter 11
Renewable Energy Technologies: Innovations and Sustainability

11.1 Introduction

Rising urgency of climate change, environmental damage, and limited supply of fossil fuels have driven a worldwide change toward renewable energy. Adoption of renewable energy sources including solar, wind, hydroelectric, and geothermal power would help the power sector, which accounts for a significant amount of world CO_2 emissions, to be mainly decarbonized. Compared to fossil fuels, these plentiful, cleaner alternatives release significantly less greenhouse gases. Changing to renewable energy will help us solve the main driver of climate change, burning of fossil fuels for generation of electricity.

Apart from lowering greenhouse gas emissions, the change to renewable energy is necessary for advancing sustainable development. Since renewable energy sources depend on constantly renewed supplies, unlike fossil fuels they are sustainable. By giving businesses, homes, and public institutions a consistent and reasonably priced energy source, which is vital, this change encourages social and economic growth. Furthermore, compared to conventional power plants, renewable energy sources may use less water; this is especially crucial in areas experiencing water scarcity.

There are obstacles to the general acceptance of renewable energy even if their obvious advantages abound. The change is hampered by economic barriers including high upfront costs and lack of finance; technology challenges like energy storage and grid integration; and policy-related obstacles include inadequate supporting legislation and industry opposition from fossil fuels. These obstacles can be surmounted, though, with suitable research and development funding, friendly legislative frameworks, and international cooperation. Adopting renewable energy helps us not only slow down global warming but also open the path for a rich and sustainable future.

H. Taherdoost, *Innovating and Integrating Emerging Technologies*, Signals and Communication Technology, https://doi.org/10.1007/978-3-031-93953-2_11

Advancement of renewable energy technology depends mostly on creativity. Not too long ago, major technological advancements improved the scalability, cost, and efficiency of renewable energy systems. Continuous innovation in materials science, manufacturing techniques, and system integration has resulted in these cost savings, which make renewable energy solutions more accessible and competitive than conventional energy sources Apart from cost savings, technological developments have resulted in more adaptable and combined energy systems that can better fit the fluctuating character of renewable energy sources, hence improving grid stability and dependability [1].

Development of renewable energy depends on sustainability in great measure. If renewable energy sources are intrinsically more sustainable than their fossil fuel equivalents, their production, deployment, and disposal can nevertheless have major environmental and social effects. By encouraging energy efficiency, waste reduction, and responsible material procurement, which helps to mitigate these effects, sustainable practices in the renewable energy sector hope to furthermore included in the idea of sustainability in renewable energy is the circular economy, in which goods and resources are recycled and reused to lower waste and environmental damage. Sustainable renewable energy sources guarantee fair distribution of the advantages of clean energy and active participation of communities in decision-making procedures, thereby fostering social fairness [2].

11.2 Overview of Renewable Energy Sources

Dealing with the worldwide threat of climate change and moving away from fossil fuels, renewable energy sources have become very vital. These sources are intrinsically sustainable since they use natural processes that are restored at either equal to or faster than their usage. Each of the primary forms of renewable energy, sun, wind, hydro, geothermal, biomass, and ocean, has unique qualities and uses. Their integration into the energy grid, storage systems, and end-use applications is shown in Fig. 11.1.

Solar energy makes heat or power from sunshine. Among the most plentiful and extensively used renewable energy sources are those photovoltaic (PV) cells, which directly convert sunlight into electricity, or concentrated solar power (CSP) systems, which utilize mirrors or lenses to focus sunlight into a small area, generating high temperatures to drive a steam turbine; both can be used to harness solar power. Adoption of solar energy has sped since PV technology's cost has dropped and solar cells' efficiency has raised. From small-scale domestic systems to massive solar farms adding significant amounts of electricity to the grid, the adaptability of solar power makes possible many uses.

Using wind turbines, the kinetic energy of wind is turned into electricity. With more effective and bigger onshore and offshore turbines resulting from technological developments, wind energy has seen explosive increase. While offshore wind farms gain from stronger and more continuous winds over the ocean, onshore wind

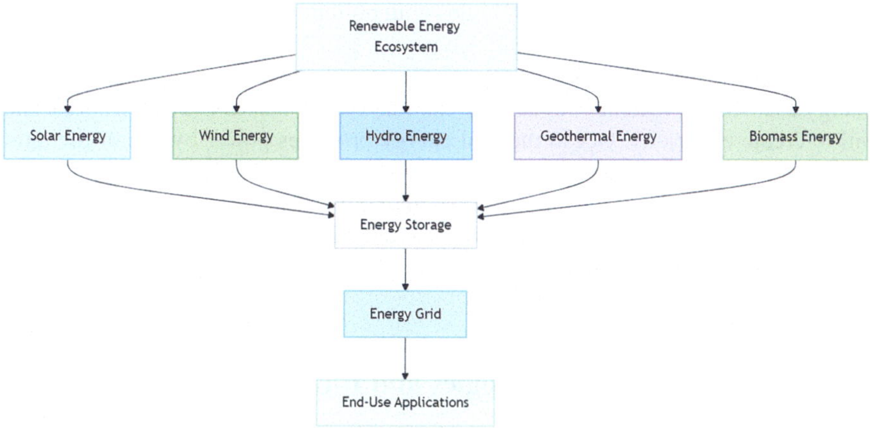

Fig. 11.1 Renewable energy ecosystem

farms usually find places with predictable wind patterns. Larger rotor blades and floating platforms for deep-water installations are two examples of turbine design innovations that have increased the possible wind energy capacity. Because wind power has cheap running costs and few pollutants during operation, it is very beneficial.

Hydro energy generates electricity from flowing water's power. Conventional hydroelectric power plants generate a reservoir via dams, then turn the gravitational potential energy of the stored water into electricity via turbines. Although run-of-river hydro systems with less environmental impact are becoming of interest, large-scale hydroelectric projects can offer notable volumes of power. Though its environmental consequences, such habitat disturbance and variations in river flow, demand careful control, hydropower's dependability and capacity to supply base-load power make it a pillar of renewable energy policies.

Using heat from the Earth's core, geothermal energy produces either direct heating or power. Underneath the surface of the Earth, geothermal power plants draw steam or hot water reservoirs from which turbines linked to power generators can be driven. Since geothermal reservoirs are not affected to weather, this energy source is quite dependable and able of supplying continuous power. Geothermal energy is location-specific, though, and the best opportunities are found in geologically active areas such Iceland, portions of the United States, and numerous East African nations.

Biomass energy is turning organic materials, such as animal and plant waste, into energy. Direct burning of biomass for heat or conversion into biofuels such as ethanol and biodiesel for use in transportation by substituting waste products that would otherwise breakdown and produce methane and replacing fossil fuels, this energy source can help lower greenhouse gas emissions. The sustainability and efficiency of biomass energy are being enhanced by developments in biomass technology including better conversion techniques and the creation of second-generation biofuels from non-food crops. Still, issues with pollution, food production, and land usage call for attention.

Ocean energy covers tidal, wave, and ocean thermal energy. Whereas wave energy uses the movement of ocean waves, tidal energy makes use of tides' rise and fall to create electricity. Ocean thermal energy produces power by use of the temperature difference between warm surface waters and chilly deep waters. Though ocean energy offers great promise, compared to other renewable sources, it is still in its early years of research. High expenses, technological complexity, and environmental issues constitute among the difficulties.

11.3 Solar Energy Technologies and Innovations

Solar energy is a renewable substitute for fossil fuels that uses the sun's energy to produce heat or electricity. Figure 11.2 shows solar energy innovations from 2010 to 2024, including thin-film solar cells, high-efficiency monocrystalline panels, perovskite cells, bifacial panels, and solar skin technologies. Photovoltaic (PV) cells and concentrated solar power (CSP) systems are the two main solar technologies. While CSP systems employ mirrors or lenses to concentrate sunlight and produce thermal energy for power generation, photovoltaic cells use semiconductor materials to convert sunlight directly into electricity.

In recent years, there have been notable breakthroughs in photovoltaic technology. The creation of bifacial solar panels, which collect sunlight from both the front and back, is one noteworthy invention. By harnessing reflected sunlight from surfaces like rooftops or the ground, this design increases energy yield. Using perovskite solar cells is another exciting advancement. Perovskites are a class of materials that combine low production costs and excellent efficiency thanks to their distinctive crystal structure. In the solar business, perovskite technology has the potential to revolutionize because of its scalability and affordability.

Utilizing mirrors or lenses, concentrated solar power systems direct sunlight onto a narrow region, producing high heat that powers a turbine to generate electricity. One of the technological innovations in CSP is the creation of molten salt storage systems, which enable energy production even in the absence of sunlight. As a thermal storage medium, molten salts are used to retain heat produced during the day and release it at night or during cloudy conditions. Utilizing heliostats, mirrors

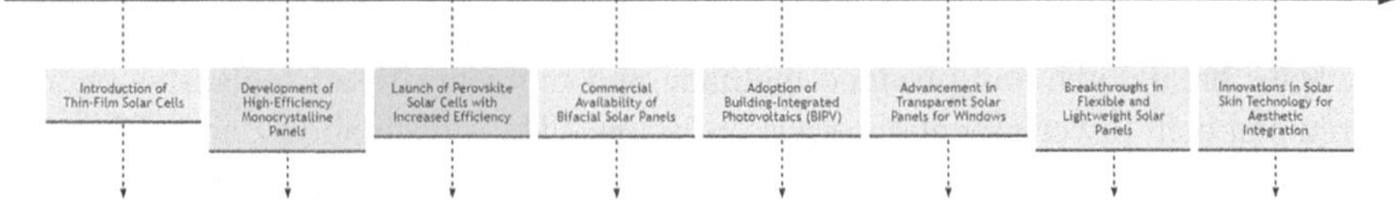

Fig. 11.2 Technological advancements in solar energy

that follow the path of the sun to maximize light concentration, is another innovation. The efficiency of CSP systems has increased recently due to advancements in heliostat design, which has decreased the price of solar electricity.

A developing trend that attempts to reduce reliance on the grid and increase building energy efficiency is the integration of solar energy into urban infrastructure. This trend is shown by building-integrated photovoltaics (BIPV), which integrate solar panels into structural elements including windows, facades, and roofs. Because BIPV systems can be made to mix in with architectural features while producing renewable energy, they have advantages that are both functional and aesthetically pleasing. Advancements in solar thermal technology have increased the energy efficiency of both residential and commercial structures. Examples of these inventions are solar air collectors and solar water heaters.

There are still a number of obstacles facing solar energy technology despite these developments. Even though solar panel efficiency is rising, problems like material deterioration and energy conversion losses still need to be addressed. Because the disposal of old or broken solar panels can have an adverse effect on the environment, recycling solar panels is also an important topic. Creating efficient PV material recycling procedures is crucial to the long-term viability of solar energy. Because solar energy is erratic due to weather patterns and the day-night cycle, innovations in energy storage technologies are required to guarantee a steady supply. Table 11.1 provides an overview of recent innovations in solar panel technology.

11.4 Wind Energy: Turbines and Advancements

The kinetic energy of wind is captured by wind turbines, which have experienced substantial technological developments in recent decades, to produce electricity. Onshore and offshore wind turbines are the two main varieties; each has specific benefits and particular difficulties. The conventional workhorse of the wind energy

Table 11.1 Recent innovations in solar panel technology

Innovation	Key features	Cost	Applications	Durability
Perovskite solar cells	Uses perovskite materials for light absorption	Moderate	Residential, commercial, research	Moderate, improving
Bifacial solar panels	Captures sunlight from both sides	High	Rooftops, solar farms, large-scale projects	High
Building-integrated photovoltaics (BIPV)	Integrated into building materials	Very high	Buildings, facades, roofs	Moderate to high
Flexible solar panels	Lightweight and bendable	Moderate	Portable devices, unconventional surfaces	Variable, generally lower
Transparent solar panels	Allows light to pass through	High	Windows, glass facades	Moderate

industry has been the onshore turbine, which is typically seen in land-based wind farms. However, more recent developments have produced more effective designs that lower the cost per megawatt-hour (MWh) and boost energy capture, such as bigger rotor diameters and taller towers. Since more power can be produced at lower wind speeds thanks to the larger scale of contemporary turbines, wind energy is now more competitive with other energy sources.

Offshore wind turbines are a rapidly developing area with a lot of promise. These turbines are placed in bodies of water to benefit from stronger and more reliable winds. They are usually larger than their onshore counterparts. Floating wind turbines are one of the innovations in offshore wind technology. They are installed on platforms that are suspended and secured to the ocean floor. Fixed-bottom turbines are impractical for deployment in deeper waters, but our technology makes it possible. By utilizing enormous offshore wind resources, floating turbines have the potential to greatly increase the geographic reach of wind energy. Improvements in materials and engineering techniques to endure challenging maritime settings have also resulted from the development of floating wind turbines.

The physical structure of wind turbines has not been left behind by technological developments in design. The operation and maintenance of wind turbines have been completely transformed by the integration of digital technologies. In order to enable predictive maintenance and decrease downtime, advanced sensors and data analytics are being employed more and more to monitor turbine performance in real time. By analyzing sensor data, machine learning algorithms may anticipate possible problems and optimize turbine operation, boosting productivity and prolonging the turbines' operating life. Furthermore, new developments in turbine control systems, like pitch and yaw control, improve the system's capacity to maximize energy output and adapt to shifting wind conditions.

Aerodynamic design advancements are part of the technology advancements in wind turbines. The efficiency of energy capture has increased with the introduction of longer and lighter blades constructed of sophisticated composite materials. Larger and more effective turbines may be designed thanks to these new materials, which not only make the blades lighter but also stronger and more durable. Through the use of computational fluid dynamics (CFD) models, the aerodynamics of blades has been optimized, resulting in the construction of blades that function well in a variety of wind conditions. The development of wind turbine designs from conventional onshore turbines to cutting-edge offshore and bladeless turbines is depicted in Fig. 11.3.

The wind energy industry still confronts a number of obstacles in spite of these developments. Grid stability is challenged by the intermittent nature of wind energy, and while developments in energy storage are assisting in mitigating this problem, further research is required to create more scalable and economically viable storage options. The installation of wind turbines necessitates a large infrastructure expenditure, particularly offshore, which may be a barrier to entrance for some places.

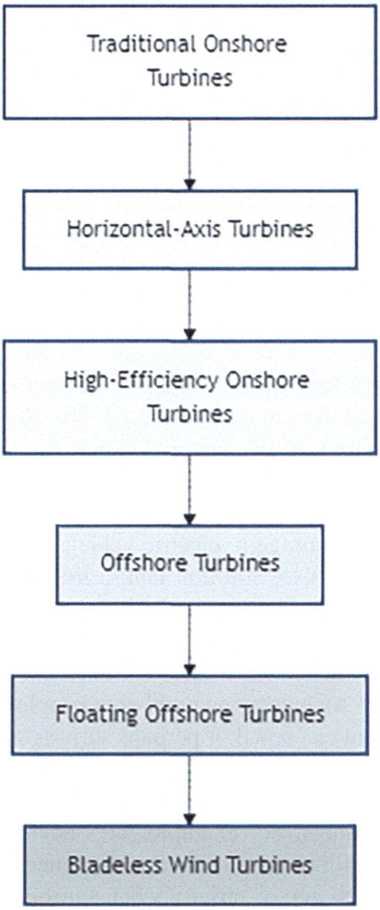

Fig. 11.3 Trend of wind turbine designs

11.5 Emerging Trends in Energy Storage

Energy storage solutions enhance grid stability and energy reliability when combined with renewable energy, as seen in Fig. 11.4. Capturing surplus energy and discharging it when required is a critical function of energy storage systems, including batteries, which guarantee a consistent and dependable power supply. This capacity smoothes out variations in energy production, which lowers the danger of blackouts and interruptions while also balancing supply and demand and improving grid stability. Energy storage technologies have advanced recently, and this is crucial for tackling the intermittent nature of renewable energy sources like wind and solar, which do not always supply power.

The creation of next-generation batteries is one of the biggest developments in energy storage. Despite being extensively utilized, conventional lithium-ion batteries have drawbacks in regard to energy density, cycle life, and safety. Researchers

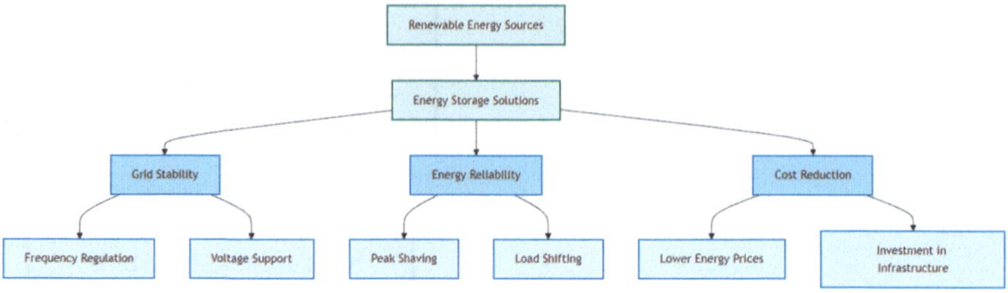

Fig. 11.4 Impact of energy storage solutions on renewable integration

are looking into different battery technologies, like flow batteries and solid-state batteries, to address these limits. Solid-state batteries have a number of benefits over liquid or gel batteries, such as increased safety, increased energy density, and extended cycle life. Solid-state batteries could greatly improve the performance of energy storage systems, making them appropriate for a range of applications, from grid storage to electric vehicles [3]. Flow batteries, which store energy in external electrolyte solution tanks, are another exciting technological advancement. Flow batteries have benefits like scalability, high cycle life, and prolonged energy discharge. The commercial viability of flow batteries is being advanced by advancements in system design and electrolyte material science [4].

An increasing number of people are interested in alternate energy storage options that can equal or perhaps surpass standard batteries, in addition to developments in battery technology. Systems for thermal energy storage (TES), which store energy as heat, are becoming more and more popular in both commercial and residential settings. For example, TES devices are used in concentrated solar power (CSP) facilities to store excess solar energy as thermal energy, which can then be converted back into electricity when sunshine is scarce. Phase-change materials (PCMs) and sophisticated insulating techniques are two recent advancements in TES materials that improve these systems' efficiency and storage capacity [5]. Similarly, long-term, large-scale energy storage using hydrogen storage has proven to be a workable option. Through the process of electrolysis, excess renewable energy can be converted to hydrogen, which can then be stored for use in combustion engines or fuel cells. To increase storage density and lower costs, research is being done on hydrogen storage materials such as chemical and metal hydrides [6].

Innovations in recycling and environmental practices are other emerging topics in energy storage. The necessity to handle the financial and environmental effects of managing storage devices at the end of their useful lives is increasing with the use of energy storage systems. The environmental impact of battery disposal is being lessened by the development of recycling technology and circular economy activities, according to recent studies. For example, improvements in lithium-ion battery recycling techniques, such as the recovery of essential elements like nickel, cobalt, and lithium, are essential for minimizing waste and lowering resource depletion. Studies on battery second-life applications, in which exhausted batteries are recycled for energy storage in less demanding uses, offer a viable way to prolong the lifespan of storage devices [7, 8].

Advanced control systems and smart grid technologies make it easier to manage energy storage resources effectively, improving supply and demand equilibrium and grid resilience. The power grid's flexibility and resilience are being enhanced by advancements in grid-scale storage systems, such as massive battery arrays and hybrid storage solutions that integrate several storage technologies [9]. These developments help to move toward a more resilient and sustainable energy infrastructure and encourage the wider use of renewable energy sources.

11.6 Sustainable Practices in Renewable Energy

As the renewable energy industry grows, it is more crucial than ever to include sustainability into its operations in order to reduce adverse effects and maximize advantages in every way. The management of supply chains is a key component of sustainable practices in renewable energy. Production of renewable energy technology, such as wind turbines and solar panels, requires intricate supply chains containing a wide range of raw materials, including metals and rare earth elements. These materials are frequently sourced from areas where there are serious issues with human rights and the environment. Sustainable supply chain methods guarantee ethical sourcing and lessen the environmental impact of material extraction and processing in an effort to address these problems. Table 11.2 summarizes various sustainability metrics for renewable energy technologies.

For example, using recycled materials to make solar panels can lower the need for virgin raw materials, which lowers the impact of mining and processing on the environment. Parallel to this, improvements in wind turbine component design have resulted in the creation of recyclable blades, which lessen the amount of waste

Table 11.2 Sustainability metrics for renewable energy technologies

Technology	Environmental impact	Resource use	Recycling/disposal
Solar photovoltaics	Low	High (silicon, rare metals)	Increasing, ongoing research on efficiency
Wind turbines	Low	Moderate (steel, rare earths)	Well-established, recycling programs in place
Hydropower	Moderate	Low (natural resources)	Limited, ongoing improvements in fish-friendly practices
Geothermal energy	Low	Moderate (land use, materials)	High, sustainable with proper management
Biomass	Moderate	Moderate (land, water)	Partial, some components can be recycled, but varies
Hydrogen storage	Moderate	High (energy intensive)	Developing, focus on efficiency and cost reduction

produced at the end of the turbine's life cycle. The renewable energy sector may greatly reduce its ecological footprint while advancing resource efficiency and sustainability by implementing the concepts of the circular economy.

A key component of improving sustainability in the renewable energy industry is the circular economy paradigm. In order to close the loop of product lifecycles through increased resource efficiency and waste reduction, this model highlights the significance of designing products with their full lifecycle in mind [10]. Designing for disassembly, recycling, and reuse is relevant to renewable energy technology.

For instance, the solar photovoltaic sector is progressively implementing procedures that center on recycling outdated panels in order to extract precious elements such as cadmium, silver, and silicon. This keeps electronic trash from building up and lessens the requirement for fresh raw materials. Similar to this, developments in wind turbine blade recycling technologies are tackling the problems related to disposing of the blades, which have been a major worry because of the amount of trash produced.

Sustainable methods in renewable energy go beyond environmental concerns to include social sustainability. This entails making certain that the advantages of developing renewable energy are dispersed fairly among communities and that the local populace participates actively in and gains from energy initiatives. Community empowerment and engagement are critical to the effective execution of renewable energy projects.

Numerous instances have demonstrated that renewable energy initiatives offer financial advantages to nearby communities, including the generation of jobs and enhanced infrastructure. For instance, solar energy projects can boost local economies through a variety of support services, and wind farms can provide jobs throughout both the construction and operation phases. But it is crucial to deal with possible social issues, such making sure that local people have a say in decision-making processes and aren't negatively impacted by large-scale initiatives.

Incentives and legislative frameworks that are effective are essential for encouraging sustainable behaviors in the renewable energy industry. Through the adoption of policies that support sustainability, governments and regulatory agencies have a significant impact on the direction that renewable energy production takes. To ensure that sustainability is incorporated into the development and disposal of renewable energy technology, for instance, laws requiring recycling and the use of eco-friendly materials might spur innovation [11].

Reward programs like grants, tax breaks, and subsidies for sustainable technology research and development help hasten the adoption of best practices in renewable energy. A more robust and sustainable energy system is facilitated by policies that support grid integration of renewable energy sources and stimulate the deployment of energy storage systems.

11.7 Conclusion

The field of renewable energy technology has experienced significant changes in the last few years due to innovation as well as increased attention to sustainability. This chapter has examined the developments in a number of renewable energy sources, such as wind and solar power, as well as the latest developments in energy storage. It has also emphasized how crucial sustainable practices are to guaranteeing that the expansion of renewable energy is socially and environmentally responsible.

It is evident from a summary of the advancements in renewable energy technology that the sector is moving quickly forward. With the advent of more effective photovoltaic cells, such as perovskite and bifacial panels, as well as integrated systems like building-integrated photovoltaics (BIPV), solar energy has significantly improved. With advancements like bladeless designs and floating turbines, which promise to improve efficiency and lessen environmental effect, wind energy technologies are also progressing. These technological advancements are essential for improving the efficacy and scalability of renewable energy sources.

With energy storage, the intermittent problems with solar and wind power can be resolved. Energy storage has become an essential part of the renewable energy scene. Advances in battery technology, like flow and solid-state batteries, are opening the door to more effective and durable storage options. Furthermore, developments in thermal and hydrogen storage are adding to the range of energy storage options, making them more resilient and versatile. Nonetheless, there are still issues with cutting costs and enhancing the sustainability and efficiency of various storage solutions.

To make sure that the development of renewable energy technology does not compromise social or environmental well-being, sustainable practices are crucial. As part of this, embracing circular economy ideas can help reduce waste and improve the way energy systems are lifecycle managed. Ethics and sustainable supply chain management must be incorporated into all phases of technology development and implementation. Furthermore, it is impossible to overestimate the importance of incentives and policies in encouraging sustainable activities. Strong legislative frameworks are necessary to encourage the use of renewable energy sources while guaranteeing that they complement more general sustainability objectives.

Renewable energy has a bright future full of possibilities for advancement and innovation. For renewable energy technologies to overcome current obstacles and realize future possibilities, research and development must continue. Governments, corporations, and communities are among the stakeholders who must work together to promote the use of renewable energy and ensure sustainability in all facets of its implementation.

References

1. Moe, E.: Renewable energy transformation or fossil fuel backlash: vested interests in the political economy. Springer, Cham (2016)
2. Byrne, J., et al.: A solar city strategy applied to six municipalities: integrating market, finance, and policy factors for infrastructure-scale photovoltaic development in Amsterdam, London, Munich, New York, Seoul, and Tokyo. Wiley Interdiscip. Rev. Energy Environ. **5**(1), 68–88 (2016)
3. Tarascon, J.-M., Armand, M.: Issues and challenges facing rechargeable lithium batteries. Nature. **414**(6861), 359–367 (2001)
4. Zeng, W., et al.: Enhanced ion conductivity in conducting polymer binder for high-performance silicon anodes in advanced lithium-ion batteries. Adv. Energy Mater. **8**(11), 1702314 (2018)
5. Zalba, B., et al.: Review on thermal energy storage with phase change: materials, heat transfer analysis and applications. Appl. Therm. Eng. **23**(3), 251–283 (2003)
6. Schlapbach, L., Züttel, A.: Hydrogen-storage materials for mobile applications. Nature. **414**(6861), 353–358 (2001)
7. Cheng, M., et al.: Sustainability evaluation of second-life battery applications in grid-connected PV-battery systems. J. Power Sources. **550**, 232132 (2022)
8. Patel, A.N., et al.: Lithium-ion battery second life: pathways, challenges and outlook. Front. Chem. **12**, 1358417 (2024)
9. Luo, X., et al.: Overview of current development in electrical energy storage technologies and the application potential in power system operation. Appl. Energy. **137**, 511–536 (2015)
10. Geissdoerfer, M., et al.: The circular economy—a new sustainability paradigm? J. Clean. Prod. **143**, 757–768 (2017)
11. Jacobson, M.Z., Delucchi, M.A.: Providing all global energy with wind, water, and solar power, Part I: technologies, energy resources, quantities and areas of infrastructure, and materials. Energy Policy. **39**(3), 1154–1169 (2011)

Index

H. Taherdoost, *Innovating and Integrating Emerging Technologies*, Signals and Communication Technology, https://doi.org/10.1007/978-3-031-93953-2

V

W

The manufacturer's authorised representative in the EU is Springer Nature Customer Service Centre GmbH, Europaplatz 3, 69115 Heidelberg, Germany. If you have any concerns regarding our products, please contact ProductSafety@springernature.com

Printed and bound by CPI Group (UK) Ltd, Croydon, CR0 4YY
30/09/2026
02225577-0007